Performance and Improvement of Green Construction Projects

Performance and Improvement of Green Construction Projects

Management Strategies and Innovations

Hwang Bon-Gang

ELSEVIER

Butterworth-Heinemann
An imprint of Elsevier

Butterworth-Heinemann is an imprint of Elsevier
The Boulevard, Langford Lane, Kidlington, Oxford OX5 1GB, United Kingdom
50 Hampshire Street, 5th Floor, Cambridge, MA 02139, United States

Notices
Knowledge and best practice in this field are constantly changing. As new research and experience broaden our understanding, changes in research methods, professional practices, or medical treatment may become necessary.

Practitioners and researchers must always rely on their own experience and knowledge in evaluating and using any information, methods, compounds, or experiments described herein. In using such information or methods they should be mindful of their own safety and the safety of others, including parties for whom they have a professional responsibility.

To the fullest extent of the law, neither the Publisher nor the authors, contributors, or editors, assume any liability for any injury and/or damage to persons or property as a matter of products liability, negligence or otherwise, or from any use or operation of any methods, products, instructions, or ideas contained in the material herein.

Library of Congress Cataloging-in-Publication Data
A catalog record for this book is available from the Library of Congress

British Library Cataloguing-in-Publication Data
A catalogue record for this book is available from the British Library

ISBN: 978-0-12-815483-0

Publisher: Matthew Deans
Acquisition Editor: Ken McCombs
Editorial Project Manager: Joshua Bayliss
Production Project Manager: Sruthi Satheesh
Cover Designer: Christian Bilbow

Typeset by SPi Global, India

Contents

List of Figures

List of Tables

Chapter 1

Introduction

1.1 RESEARCH MOTIVATION

Along with a growing recognition of the importance of sustainability among global construction communities, green construction has achieved rapid growth over recent decades. However, despite its success, green construction confronts various problems caused by poor project management approaches. Current literature shows that 30% of green construction projects experience rework (Hwang et al., 2016), 50% of green construction projects are plagued by delay (Abdel-Razzaq, 2016), and 90% of green construction projects require cost premiums of up to 21% to ensure their completion (Dwaikat and Ali, 2016). Furthermore, it has been reported in the literature that green construction projects face greater risks than conventional construction projects (Hwang et al., 2015a, 2017c), and that green construction projects pose a greater safety hazard to frontline workers than conventional ones (Dewlaney et al., 2011). The statistics appear to be sending a strong signal to the authorities, industry and academia that it is time to devote careful, close, and particular attention to the management challenges and issues in green construction projects.

The primary mission for a green construction project is to create a built structure that consumes fewer resources for construction, releases fewer greenhouse gas emissions during operation, and provides the end users with a pleasant, healthy, and comfortable living environment. However, this mission can only be accomplished after the project is successfully delivered, which in turn is mainly subject to the successful management of the project (Glavinich, 2008; Kubba, 2010). Currently, despite extensive existing literature on green construction, most of the literature concentrates on the technical aspects, such as energy saving and simulation, green architecture and design, green construction technologies and materials, and green building rating systems, and very little of it addresses the management problems that occur in green construction projects (Zuo and Zhao, 2014; Darko and Chan, 2016). Therefore, this book aims to bridge this knowledge gap by conducting a thorough investigation of the various management challenges and issues existing in green construction projects, and by exploring feasible solutions that can be applied to these challenges and issues.

Performance and Improvement of Green Construction Projects. https://doi.org/10.1016/B978-0-12-815483-0.00001-6

1.2 RESEARCH SCOPE

A green building is one that is environmentally responsible and resource efficient throughout its life-cycle (Environmental Protection Agency, 2016; Kibert, 2008). It has green and environment-friendly features that can improve the building performance and provide the end users with a good and healthy indoor environment (Yudelson, 2010a). Due to the special characteristics of green buildings, the means by which they are constructed are different compared to conventionally-built structures (Dwaikat and Ali, 2016). Thus, particular emphasis should be placed on one critical question, namely, how green construction projects can be successfully implemented. Based on a thorough investigation within the green construction industry of Singapore, this book attempts to answer this question by looking at a wide range of management topics associated with the implementation of green construction projects. First, this book reviews the practices and research efforts relating to green building rating systems. Then the obstacles that hinder the promotion of green construction in the current construction industry are investigated. In addition, the key knowledge areas and skills that are required for the successful management of green construction projects are examined. Furthermore, this book reviews the policies, practices, and major research efforts conducted in the area of green construction project financing. Moreover, this book presents a holistic view of the management strategies and innovations related to green construction projects, focusing on the assessment and improvement of management performance in the areas of cost, schedule, quality, safety, risk, productivity and leadership development.

It should be noted that the content of this book is primarily based on a series of empirical studies conducted in Singapore. Singapore has been widely viewed as a leader in advocating for sustainability in the global construction community with its efficient strategies and initiatives (WorldGBC, 2013). A large number of green construction projects have been implemented in the country recently, which provide a number of exemplary cases to generate extensive knowledge concerning the management of green construction projects. Furthermore, the author has been conducting research in the area of green construction for at least 10 years at the National University of Singapore, a world leading university for sustainability research, and thus the author is quite familiar with the past, present, and future of the green construction industry in Singapore.

1.3 RESEARCH OBJECTIVES

The primary objectives of this book are to:

1. Review prevailing green building rating systems used by the global green construction community.
2. Identify major obstacles that hinder the promotion of green construction in the current construction industry, and explore practical solutions that can overcome these obstacles.

3. Explore key knowledge areas and skills required for the successful management of green construction projects.
4. Review the practices, policies, initiatives, and research efforts relating to green construction project financing.
5. Investigate the current status of cost premiums in green construction projects, and explore solutions that can reduce cost premiums and improve cost performance.
6. Examine the schedule performance of green construction projects and propose corresponding solutions for improvement.
7. Examine the safety performance of green construction projects and come up with practical solutions for improvement.
8. Address the rework issues in green construction projects and seek a set of feasible solutions to reduce rework.
9. Examine the productivity performance of green construction projects, identify the critical productivity factors, and come up with practical solutions to improve productivity.
10. Develop a computational model to assess the risks in green construction projects.
11. Oversee the risk management of green retrofit, residential and commercial construction projects, and propose relevant risk mitigation measures.
12. Investigate leadership development during the implementation of green construction projects.

The various management challenges facing green construction projects would be well addressed if the 12 objectives listed above could be achieved. Moreover, achievement of the above objectives would provide comprehensive and innovative knowledge regarding the management of green construction projects, which could be widely shared.

1.4 RESEARCH SIGNIFICANCE

Although increasing research effort has been devoted to the area of green construction recently, there have been little research to address management problems and challenges in green construction projects (Zuo and Zhao, 2014; Darko and Chan, 2016). Consequently, there are no mature management theories that academics can refer to when they are trying to solve complicated management problems occurring in green construction projects. To address this issue, this book will describe a series of theoretical frameworks that compare the management performance of green construction projects with that of conventional projects in the areas of cost, schedule, safety, quality, risk, productivity, and leadership. Furthermore, these frameworks may also serve as a solid basis for further studies in the area of green construction, making this book beneficial to those in academia.

This book may also be helpful to policymakers as it can provide them with a clearer picture of the obstacles, challenges, risks, and barriers to implementing green construction projects. Based on the knowledge gained from this book, policy makers should be able to develop policies that are more effective in promoting and supervising implementations of green construction projects. It is necessary for policymakers to refer to the views of industry when they establish or amend a policy concerning the green construction industry. Unfortunately, there is currently a gap between policymakers and industry that hinders the development of effective policies. As the findings of this book are mainly derived from empirical results obtained from industry practitioners, this book can be a good source for policymakers to learn more about the stance of industry in implementing green construction projects, so that they in turn can develop more deliberate policies.

Although green construction has been promoted and implemented for nearly two decades, the experiences and lessons learned from this practice have been barely summarized. This book aims to fill this knowledge gap. This book highlights the numerous management issues, problems, and risks that industry practitioners may encounter in implementing green construction projects, and examines the differences between green and conventional projects in terms of management performance and approach. Thus, to a certain extent this book can serve as a helpful manual or handbook for industry practitioners working on green construction projects. Relying on the knowledge gained from this book, industry practitioners can enhance their understanding of, and capabilities in, managing green construction projects.

Owing to the rapid development of green construction, it is predictable that numerous graduates majoring in, for example, construction engineering and management, will participate in the green construction business in the near future. Thus, it is necessary and imperative to provide them with some basic knowledge of green construction (particularly regarding the challenges related to the management of green construction projects as well as the management approaches) before they commence their professional careers. This book can provide a platform in which undergraduate students may obtain some fundamental knowledge regarding the management of green construction projects. Furthermore, questions, such as, "What is a green building? What are its elements? What are the challenges in managing green construction projects? What feasible strategies can be taken to respond to these challenges?" are also systematically addressed in this book. As most information in this book is derived from empirical studies with industry practitioners, the book can also help students learn more about real-world cases and examples of the implementation of green construction projects.

1.5 STRUCTURE OF THE BOOK

This book is composed of 16 chapters as briefly outlined below.

Chapter 1 introduces the research motivation, research scope, research significance, and research objectives, as well as the structure of the book.

Chapter 2 provides background material, including a review of the concepts of green buildings and the rationales behind them. It also discusses green construction projects and their unique characteristics. A comprehensive overview of the Singapore green construction industry is also presented.

Chapter 3 focuses on the research methodology used in this book. It describes the research design and data collection methods. Various approaches for data analysis are also introduced.

Chapter 4 reviews the prevailing green building rating systems used in the world, in terms of their timeline, particular versions, key evaluation criteria, and the hierarchy of rating results. Furthermore, it summarizes the major research efforts conducted in this area, as well as the future research directions.

Chapter 5 examines the major obstacles impeding green construction. It also proposes practical solutions to overcome those obstacles.

Chapter 6 summarizes the critical challenges that industry practitioners may encounter during their implementation of green construction projects. It also explores the key knowledge areas and skills required for the management of green construction projects.

Chapter 7 reviews the practices, policies, initiatives, and research efforts relating to green construction project financing. It also discusses the future research directions in this area.

Chapter 8 investigates the current state of cost premiums in green construction projects, and compares it to conventional construction projects. It also explores solutions that can reduce cost premium and improve the cost performance of green construction projects.

Chapter 9 examines the schedule performance of green construction projects. This chapter also explores the critical factors affecting schedule performance of green construction projects and compares them with those of conventional construction projects. In addition, it proposes solutions to improve the schedule performance of green construction projects.

Chapter 10 examines the safety performance and safety issues of green construction projects and compares them to those of conventional projects. Solutions for improving safety are also discussed.

Chapter 11 addresses the issue of rework in green construction projects. It examines the current state of rework in green construction projects and compares it to that of conventional construction projects. Critical rework factors are identified and compared between green and conventional construction projects. Solutions to reduce rework in green construction projects are also proposed.

Chapter 12 examines the productivity performance of green construction projects. It also investigates the critical productivity factors of green construction projects and makes a comparison with conventional construction projects. In addition, this chapter provides solutions to improve the productivity performance of green construction projects.

Chapter 13 is dedicated to the development of a computational model to assess the risks of green construction projects. It also explores the top 10 risk factors and top 3 risk groups for green construction projects.

Chapter 14 examines the risks for green retrofit of residential and commercial construction projects. It investigates the criticalities of these risks and compares them with those of conventional construction projects. Furthermore, the relevant risk mitigation measures for green construction projects are provided.

Chapter 15 explores the leadership characteristics and styles of project managers on green construction projects. Based on the diverse leadership characteristics, it also identifies two leadership groupings that prevail in current green construction projects.

Chapter 16 summarizes the main findings of the book. In this chapter, the major contributions of the book both to knowledge and practice are presented. This chapter also discusses the limitations of the book and proposes recommendations for future research work.

Chapter 2

Background

2.1 GREEN CONSTRUCTION PROJECTS

2.1.1 Green Buildings

There has been a growing concern about global climate change as a result of increasing greenhouse gas (GHG) emissions (Hwang et al., 2015a; Wu et al., 2014a; Yuan et al., 2015). If no measures are taken to reduce GHG emissions, at least 5% of global GDP will be lost because of the adverse influence of climate change (Wu et al., 2015; Stern, 2007).

As one of the largest sources of GHG emissions, the building and construction industry is facing increasing pressure to reduce them (Wu et al., 2014b; Zuo et al., 2015). The World Green Building Council (WorldGBC, 2013) reported that buildings are responsible for one-third of GHG emissions in the world. Moreover, previous studies have shown that the building and construction industry is a big energy consumer that has consumed 40%–50% of global energy and 40% of global raw materials, and released 40% of the waste disposed of in landfills (Yang and Zou, 2014; Yang et al., 2016; United Nations Environment Programme (UNEP), 2011). However, this indicates that the building sector has the greatest potential for delivering significant cuts in GHG emissions at low or no cost, or, indeed, at a net savings to economies. In recent years, there has been an increasing focus on sustainability and an apparent shift toward green buildings and green construction all over the world (McGraw-Hill Construction, 2013; Qin et al., 2016; Zuo and Zhao, 2014).

The development of green building represents a major and cost-effective way to reduce GHG emissions and is sweeping across most of the world. The definition of "green building" varies from different construction perspectives. Generally, the goal of green buildings is to achieve energy and resource efficiency, realizing long-term economic, environmental and social health (Yoon and Lee, 2003; Sahamir and Zakaria, 2013). According to the US Green Building Council (United States Green Building Council (USGBC), 2007), a green building is a building that has been carefully designed, constructed, and operated to boost environmental, economic, and productivity performance, as compared to conventional buildings. To put it in simpler terms, green buildings are designed and built in such a way that they use less energy and fewer resources than conventional buildings and attempt to minimize the impact on

Performance and Improvement of Green Construction Projects. https://doi.org/10.1016/B978-0-12-815483-0.00002-8

the environment (Yudelson, 2010a). Green buildings are healthy facilities that are designed and built in a resource-efficient manner using ecologically based principles (Kibert, 2008).

2.1.2 Rationales Behind Green Buildings

Green buildings have environmental, economic and social benefits (Yuan and Zuo, 2013; Zuo and Zhao, 2014). First, the building sector would achieve a great and efficient reduction of GHG emissions if appropriate green technologies, materials, and construction methods were used. Wedding and Crawford-Brown (2008) found that, compared to green buildings, nongreen buildings experienced 50% higher energy consumption, 100% greater outdoor water use, and 30% higher indoor water consumption. Fowler and Rauch (2008) reported that some green commercial buildings consumed 26% less energy and saved 13% in maintenance costs compared to traditional commercial buildings. Carpenter (2009) found that the average energy savings of green building projects were up to 40% and that the long-term energy savings could be higher "once the first-year operational issues are worked out."

In addition to saving energy, the comfortable environment of a green building can increase the occupants' satisfaction, positively influencing the health and productivity of occupants and representing social benefits (Asdrubali et al., 2013). Tollin (2011) found that certified green buildings would provide healthier environments for work and play. This could potentially result in lower absenteeism and higher productivity rates among employees. Singh et al. (2010) and Thatcher and Milner (2014) both investigated the effects of a green office building on the perceived health and productivity of the occupants and found that the green building significantly contributed to the increased self-reported productivity and physical well-being of employees. Also, Barrett et al. (2013) carried out a study on 751 pupils from 34 varied classrooms in seven different schools in the United Kingdom. The results showed that the "best" classrooms, so defined by six significant built environment design parameters (color, choice, connection, complexity, flexibility and light) were estimated to have an impact that equates to the typical progress of a pupil over 1 year.

Moreover, there are also tangible benefits from green buildings. In the United States, compared with traditional buildings, green buildings can reduce operating cost by 8%–9%, increase building value by 7.5%, increase occupancy ratio by 3.5%, and improve return on investment by 6.6% (Ashuri and Durmus-Pedini, 2010). Because of these social benefits, green buildings further have the potential to generate higher rent and sale prices. A study carried out in Hong Kong indicated that "green development" is one of the considerations in the purchasing of apartments. Additionally, end-users are willing to pay more to purchase apartments with green features (Chan et al., 2009).

2.1.3 Green Construction Projects

The "construction of green buildings" and "sustainable construction" are sometimes used interchangeably. However, the construction of green buildings is only part of sustainable construction. According to Kibert (2008), sustainable construction addresses the ecological, social, and economic issues of a building in the context of its community. Sustainable construction is applied throughout the life cycle of construction, from preconstruction to disposal of the building. Such construction aims to reduce the impact of construction practice on the environment through the planning and management of a project complying with the contract document (Glavinich, 2008).

Green construction has attracted more attention in recent years, and there has been an apparent shift toward green construction (Hwang et al., 2015d). Green construction has evolved to become a necessity for environmentally conscious industry professionals, owners, developers, government officials and the remaining stakeholders (Ashuri and Durmus-Pedini, 2010; Shi et al., 2012; Wu and Low, 2014). Higher energy prices, increased building material costs, and legislation and regulatory incentives, are also pushing the green construction market to grow and expand. Thus, environmentally green construction has experienced significant growth in the past 10 years (Lim et al., 2015; Zuo et al., 2013).

2.2 OVERVIEW OF SINGAPORE GREEN CONSTRUCTION INDUSTRY

2.2.1 Singapore Construction Industry

Singapore is a city-state with limited land area and natural resources, both of which have made sustainability a necessity, rather than an option, to the country (Hwang et al., 2015d). Over the past five decades, Singapore has been struggling to achieve sustainability in its various industries (Ministry of the Environment and Water Resources and Ministry of National Development, 2014) and the building and construction industry is one of its primary emphases.

In Singapore, the construction industry is a key contributor to economic growth. According to a report from Ministry of Trade and Industry Singapore (MTI, 2015), the construction industry grew approximately 3% in 2014, which was higher than the overall rate of economic growth. The Building and Construction Authority (BCA) of Singapore reported that Singapore's construction output was about 37.7 billion Singapore dollars (SGD) in 2014 and that it would be 26–37 billion SGD from 2015 to 2018 (BCA, 2015b). The adoption of green building and the movement toward sustainability is an approach to gain a competitive edge in environmental/ecological technology and to export services to developing countries (MTI, 2016).

2.2.2 Green Schemes and Legislations in Singapore

With increasing global concern about the environment, Singapore has shifted its focus to making sustainable development a national priority (Hwang et al., 2015b). It has been recognized as a leader in advocating sustainability in the building and construction industry with its up-to-date and efficient green strategies and initiatives (WorldGBC, 2013). Singapore has embarked on a mission to make construction more environmentally friendly by launching a series of Green Mark Schemes and Green Building Masterplans to improve the industry.

In January 2005, BCA introduced the Green Mark Scheme (GMS) to aid in the greening of Singapore's current and future buildings. To intensify the green building movement, BCA launched the first and second Green Building Masterplan in 2006 and 2009, respectively. The key target of the second Green Building Masterplan is to have at least 80% of the buildings in Singapore green by 2030 (BCA, 2009). To achieve this target, the Singapore government has also launched a series of incentive plans to encourage developers, building owners, and project consultants to adopt environmentally-friendly design, technologies, and practices in their construction projects (BCA, 2015g,h). Examples include the Green Mark Incentive Scheme for new buildings introduced in 2006, the Green Mark Incentive Scheme for existing buildings introduced in 2009, and the retrofitting scheme to encourage existing buildings to go green and be more energy efficient as well as requiring new buildings above $2000\,m^2$ to achieve Green Mark certification (BCA, 2013c).

The key initiatives of the second Green Building Masterplan include getting the public sector to take the lead in achieving Green Mark (GM) certification, encouraging the private sector to attain higher GM ratings through the provision of additional gross floor area, stepping up the development of green building technologies through more Research & Development and pilot projects, building industry capability through training, profiling of Singapore as a hub for green building development, and imposing minimum standards (BCA, 2009).

Stimulated by this comprehensive suite of policies and initiatives, the green building and construction industry in Singapore has achieved rapid development. The number of green buildings increased from 17 in 2005 to 2300 in 2015, accounting for about 27% of Singapore's total gross floor area (BCA, 2014a, 2015c). In light of the long way to the vision of 2030, BCA further introduced the third Green Building Masterplan to accelerate the green agenda (BCA, 2014a). The third Green Building Masterplan highlights building capability in the industry, engaging the tenants and occupants in a closer partnership between the people and the private and public sectors, driving consumption behavioral adjustments, as well as developing an environment that addresses the well-being of the people (BCA, 2014a). Furthermore, three strategic goals, namely "Continued Leadership," "Wider Collaboration and Engagement," and "Prove Sustainability Performance," were included in the green agenda to productively guide future initiatives toward its vision and goal (BCA, 2014a).

Furthermore, the Energy Conservation Act, which mandates energy management practices for large energy users, such as the construction industry, has been passed by the Singapore government, posing some potential challenges in delivering and managing green construction projects for both new and existing facilities. Also, regulations were enacted in 2008 to make the Green Mark mandatory in the construction industry. With that, industry stakeholders began to recognize that going green had shifted from being a choice to being an obligation. The drive toward more buildings being Green Mark certified has grown beyond Singapore to reach overseas in such countries as Thailand, Vietnam, Indonesia, and China (Hwang and Tan, 2012).

2.3 CHARACTERISTICS OF GREEN CONSTRUCTION PROJECTS

Green building is the creation of structures that are environmentally responsible and resource-efficient throughout a building's life-cycle (Environmental Protection Agency, 2016). It includes elements that are energy and water efficient, are environmentally friendly, that provide a good indoor environmental quality, and that have green features to improve building performance (BCA, 2016a,c). Due to the unique characteristics of green buildings, the means by which they are constructed are different as compared to traditional building projects (Dwaikat and Ali, 2016).

2.3.1 Delivery Systems

A majority of today's construction projects are still carried out in accordance with traditional methods and norms, in which short-term solutions are favored over long-term ones, with material, technical solutions, and managerial approaches that can seldom be classified as innovative green technology and practice (Demaid and Quintas, 2006; Gluch et al., 2009). Especially, concerning delivery systems, design-bid-build, construction management-at-risk, and design-build are commonly adopted for conventional construction projects, and each method has pros and cons.

To successfully deliver green buildings, specific modifications to traditional project management processes and practices are needed (Robichaud and Anantatmula, 2011). For example, as a design having a significant impact on the cost, the special design factors mentioned above need to be considered early in the design stage, which affects the project delivery method. Different types of delivery methods, such as the most frequently used design-bid-build (DBB) and the increasingly popular design-build (DB) model, should be first considered for green buildings (CMAA, 2012). Empirical research has shown that DB was used for 75% of green building construction projects (Molenaar et al., 2009).

In addition, DB and construction management at risk (CMR), have a better chance to provide a high level of integration by facilitating the early involvement of contractors (Mollaoglu-Korkmaz et al., 2013). Expert inputs from the contractor can help avoid changes in design and reworks in the later stages of the project, which can cause a major delay in the project schedule and incur extra cost. Moreover, this delivery system ensures that the end product is consistent with the design, which is especially important for green building construction in which various specifications have to be met before it can be certified as a green building. A green building project also requires team building at the start of the project, which should include all relevant personnel involved in the project. Specialists with expertise in green building, such as building energy performance and green building certification, should also be engaged. This is because these people have a deeper comprehension of the concept of green building and are familiar with the standards and requirements of green certification.

2.3.2 Contract

During green building construction procurement, particular attention has to be placed on green requirements, which are typically found in the specifications of the contract document (Glavinich, 2008). Such requirements usually specify the types of material and equipment to be used and ensure that they satisfy the minimum standard for environmental sustainability. Additionally, sustainable construction practices and measures are another type of green specification addressed in project specifications. These sustainable construction practices and measures can be found in the LEED Green Building Rating System for new construction, which includes measures such as construction activity pollution prevention and conservation of existing natural areas (United States Green Building Council (USGBC), 2009). Such detailed specifications may not be needed in a typical construction contract. Before contracting with contractors and other professionals, such as project managers and consultants, the developer selects those who are experienced in the field of green building and sustainable construction (Addis and Talbot, 2001). In Singapore, developers can engage personnel who are Certified Green Mark Managers (GMM) or Green Mark Professional (GMP) (BCA, 2016e). Similarly, in the United States., developers can engage LEED Accredited Professionals (LEED-AP) who have passed the USGBC training and testing program on green building principles (Kibert, 2008; United States Green Building Council (USGBC), 2009). A Green Request for Proposal can be declared to recruit an architect and construction manager. The Green Request for Proposal assures the developer that the professionals possess a comprehensive understanding of green building concepts. Such an understanding allows the professionals to appreciate the aims and objectives of a green building construction project so that the project may be executed efficiently. In a conventional building contract, the contractors are often selected based on the lowest cost tendered, but may not have the ideal qualities required for that particular project.

2.3.3 Design

First, the design of a green building is different from that of a traditional building because designers need to adopt an integrated design approach. Designers should understand interactions among the various building systems and interactions between building systems and the environment so as to allow each system to be optimized (BCA, 2010a). Unique design factors, such as the building envelope, space orientation, and building materials, need specific consideration. The building envelope limits a number of external forces, such as the daylight and temperature, by shielding the interior of a building from the exterior environment (BCA, 2010a). Space orientation aims to reduce some external forces by considering the direction of sun travel (BCA, 2010a). Moreover, sustainable building construction materials are considered and crafted to precise specifications so as to ensure that a building is environment friendly (BCA, 2013b). These design factors are thought of as best practice in a traditional building design, but not a necessity.

2.3.4 Construction

A green building construction project values the efficient use of resources in its construction as well as in the use of the building (Dwaikat and Ali, 2016). In contrast to a traditional building construction, a green building construction usually must follow an additional set of requirements to help the building gain a green recognition.

One example of such practices is a waste management plan (Addis and Talbot, 2001) to minimize waste generation at the construction site (Kibert, 2008). A green building construction also has to adopt sustainable practices, such as using recycled aggregates for concrete work and using timber from renewable sources (Bill and Roger, 2001). To protect the environment, the use of sustainable construction practices, such as the use of recycled concrete aggregates, can help a building gain points for the consideration of Green Mark (BCA, 2013b). In Leadership in Energy and Environmental Design (LEED), considerations have also been made for waste management in the storing and collection of recyclables, as well as having regard for reasonably good air quality during the construction phase to promote worker health (United States Green Building Council (USGBC), 2009).

Also, the main contractor and project manager have to ensure that pollution from the construction is kept to a minimum by controlling soil erosion, waterway sedimentation and airborne dust generation (USGBC, 2009). Furthermore, the natural habitat should be conserved through careful siting of the building to minimize any disturbance to the existing natural environment. These considerations are often neglected in traditional construction.

2.3.5 Commissioning and Closing Out

The commissioning and closing out of a green building project is usually more complicated than that of a conventional project. This is particularly true when the developer wishes to attain third party green certification (Glavinich, 2008), such as LEED, BREEAM, and Green Mark. There is also a responsibility to impart knowledge of the green building systems to new facilities management teams and end users to maintain sustainability (Addis and Talbot, 2001). In addition, ease of maintenance has to be ensured.

In light of the above, research on the project management and risk management of green construction projects is necessary. Moreover, specific modifications to traditional project management processes and practices are needed.

2.4 SUMMARY

This chapter has reviewed the concepts of green buildings as well as the rationales behind them. This chapter has also discussed green construction projects and their unique characteristics in terms of delivery systems, contract, design, construction, and commissioning and closing out. Comprehensive overviews of the Singapore construction industry and the green schemes and legislation implemented in Singapore have also been presented.

Chapter 3

Methodology

3.1 RESEARCH DESIGN

To ensure the achievement of diverse research objectives, rigorous research design must be established as a guide for the entire research process. According to Creswell (2013), research design refers to the logically-linked plans and procedures that address the topic under investigation. Research design mainly considers four aspects of the research: the questions to study, the relevant data, the data collection methods, and the analysis of gathered data. A proper research design depends on the issue being addressed, the personal experiences of the researcher, and the target audience of the survey.

To carry out the diverse procedures included in the research design, research methods are needed. Research methods can be qualitative, quantitative, or mixed. Qualitative methods ask open-ended questions and use interview, observation, and document data that involve text and image analysis (Maxwell, 2012; Patton, 2005). Quantitative methods ask instrument-based questions, use performance and attitude data, and conduct statistical analysis and interpretation (Creswell, 2013; Bernard, 2012). As for the mixed method, it combines the qualitative and quantitative methods, and allows triangulation of data sources to determine convergence of the qualitative and quantitative methods used (Kothari, 2004; Creswell, 2013). This book looks into and addresses the various management problems occurring in green construction projects, which are varied, complicated, and difficult to handle by a single research method. Therefore, this book has adopted a mixed method approach that combines both qualitative and quantitative methods.

Focusing on various management problems in green construction projects, a unified and stepwise research design is adopted, as illustrated in Fig. 3.1. Step 1, a comprehensive literature review, is conducted first to understand the various management problems and to seek relevant solutions. Step 2, a questionnaire, is developed based on literature review and then disseminated widely to gather industry practitioners' perceptions of the management problems and solutions. The questionnaire is verified by an industry expert before being sent out. In Step 3, data collected from the questionnaire survey are analyzed by diverse approaches, such as parametric and nonparametric statistical tests and fuzzy synthetic evaluation. Step 4, postsurvey interviews, are conducted with industry experts to validate the findings obtained from the questionnaire survey.

Performance and Improvement of Green Construction Projects. https://doi.org/10.1016/B978-0-12-815483-0.00003-X

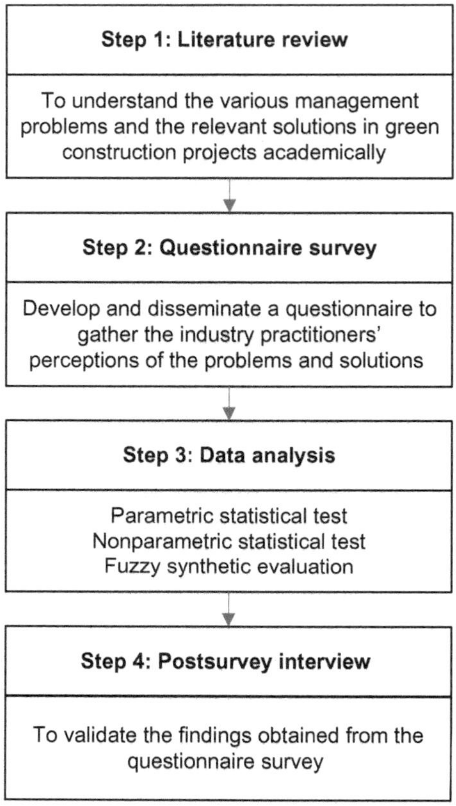

FIG. 3.1 Research design of the book.

3.2 LITERATURE REVIEW

Literature reviews are a basis for research in nearly every academic field. They are conducted as the very first step when addressing the various management problems in green construction projects. In this book, the main purpose of conducting a literature review is to fully understand the state of management problems in green construction projects, and to spot the relevant solutions that can tackle these problems. The sources for a comprehensive literature review commonly include reputable peer-reviewed journals in the areas of construction and sustainability, conference papers, reports and books relevant to green building, and news from the internet and newspapers.

3.3 DATA COLLECTION

3.3.1 Questionnaire Survey

As a systematic method of collecting data based on a sample, the questionnaire survey technique has been widely used to obtain professional views in sustainable construction research (such as Hwang and Ng, 2013; Wu and Low, 2012; Shi et al., 2012; Liu et al., 2012). Thus, the questionnaire survey is conducted to solicit opinions from experts about their perceptions of green construction projects. For each study, a questionnaire was first developed based on a comprehensive literature review. Subsequently, a two-step process was adopted to test the validity and relevance of the questionnaire. The questionnaire was first reviewed by an expert on question construction, ensuring that the survey did not contain common errors, such as misleading, confusing or double-barreled questions. Then, a presurvey was conducted through face-to-face interviews with a few construction industry professionals who had several years of experience in the local construction industry, especially in green building construction projects. After receiving feedback from them, relevant changes were made to finalize the questionnaire.

In the final survey questionnaire, before the main body, an introduction letter provided relevant definitions as well as the objectives. In the main body, the questionnaires included questions meant to profile the firms and respondents. Additionally, for each chapter, the research questions were presented in the questionnaire, and the respondents were requested to give their opinions. For instance, to assess the criticality of factors affecting the productivity of green construction projects, the respondents were requested to assess the likelihood and impact of each risk in conventional and green projects, respectively. In the questionnaire, the respondents were commonly asked to assess the research problems using a five-point Likert scale.

3.3.2 Interviews

Although the questionnaire is considered as the most cost-effective and most popular way to collect data relating to attitudes, opinions, and behaviors (Gravetter and Forzano, 2012), survey responses could be subjective, and thus validation and caution in generalizing the results are necessary (Hwang and Yang, 2014). Therefore, postsurvey interviews were carried out with several professionals to collect the rationale behind the survey results. The professionals were selected based on their willingness to participate in the interviews, and this strategy was adopted by Wilkins (2011).

3.3.3 Survey Population

The population of these research studies mostly consisted of professionals who were from companies under the BCA directory of registered contractors and licensed builders, or professionals who were from firms under the Singapore Institute of Surveyors and Valuers (SISV) or project managers with the relevant Green Mark Certification, such as Green Mark Managers (GMM) and Green Mark Professionals (GMP). To be certified GMMs and GMPs, the applicants should possess at least 3 or 5 years of relevant local experience, and complete at least one to three green building projects in Singapore (BCA, 2016e).

The professional capability and industry experience of the targeted respondents ensured constructive and reliable inputs to the survey.

The questionnaires were randomly sent out to gather responses from these professionals, who were working as developers, contractors, and consultants. Validation of the completed questionnaire was performed before any statistical analysis. Moreover, if the sample size was not less than 30, statistical analysis could be carried out according to the generally accepted rule (Hwang et al., 2015a; Ott and Longnecker, 2010; Zhao et al., 2013a).

3.4 DATA ANALYSIS

3.4.1 Statistical Analysis Methods

This book frequently adopts the Likert scale to scaling responses in questionnaires. Likert scale data have been considered as ordinal scale data and have distinct characteristics, such as discrete instead of continuous values and restricted range. Clason and Dormody (1994) investigated the use of Likert scales in international journals and found that a greater number of papers adopted parametric statistical methods. These parametric methods included, but are not limited to, the t-test, the analysis of variance F-test (Clason and Dormody, 1994), and Pearson correlation (Choi et al., 2010). Moreover, some previous studies argued that the flexibility and power of the parametric methods can outweigh the small biases that they may entail (Allan, 1976; O'Brien, 1979), and interpretations of results from parametric methods might be easier and provide more information (Allen and Seaman, 2007). In addition, Norman (2010) argued that parametric statistics can be utilized to analyze Likert data with unequal variances and nonnormal distributions, without fear of drawing the wrong conclusions. Furthermore, de Winter and Dodou (2010) concluded that the parametric and nonparametric statistical methods for five-point Likert scales are equally powerful, especially when the sample size is no less than 30. Hence, there is no need to replace parametric statistical tests by nonparametric methods when the scale of measurement is ordinal and not interval (Zumbo and Zimmerman, 1993). In light of the above, the studies involved in this book adopted the parametric methods, including descriptive means, one sample t-test, paired t-test, and Spearman rank correlation.

3.4.1.1 Cronbach Alpha Coefficient

As the questionnaire survey uses the five-point scale to solicit opinions from experts, the Cronbach alpha coefficient (α) is used to measure the internal consistency of the data obtained from the five-point rating scale. The value of α ranges between 0 and 1, where 1 indicates the highest reliability and validity of the quantitative inputs (Santos, 1999). Normally, the data achieves internal consistency if α is greater than 0.7.

3.4.1.2 *One Sample* t-*Test*

The one-sample *t*-test is a statistical method widely used to determine whether a sample comes from a population with a specific mean, which can be hypothesized as needed (e.g., Hwang et al., 2016, 2017c; Liu et al., 2016; Ning et al., 2013). The null hypothesis H_0 is that the mean score is not statistically different from the hypothesized value, while the alternative hypothesis H_1 is that the mean score is statistically different from the hypothesized value. Normally, the one sample *t*-test is performed at a confidence level of 95% with a *P*-value of .05. If a *P*-value is less than .05, the null hypothesis should be rejected, suggesting that significant difference exists between the sample mean and the hypothesized value.

3.4.1.3 *Paired Samples* t-*Test*

The paired samples *t*-test, sometimes called the dependent sample *t*-test, paired *t*-test, or repeated measures *t*-test, is a statistical procedure used to determine whether the mean difference between two sets of observations is zero (Statistics Solutions, 2017). In a paired sample *t*-test, each subject or entity is measured twice, resulting in pairs of observations (Norušis, 2006). Common applications of the paired samples *t*-test include case-control studies or repeated-measures designs (Shan et al., 2016; Hwang et al., 2013b). Thus, this book employs the paired samples *t*-test to check if there are significant differences between the industry practitioners' perceptions of green and conventional construction projects. The null hypothesis for paired samples *t*-test is that the true mean difference between the paired samples is zero, while the alternative hypothesis assumes that the true mean difference between the paired samples is not equal to zero. Normally, the paired samples *t*-test is performed at a confidence level of 95% with a *P*-value of .05. If a *P*-value is less than .05, the null hypothesis should be rejected, suggesting that significant difference exists between the paired samples.

3.4.1.4 *Spearman Rank Correlation Coefficient*

The Spearman rank correlation coefficient (*S*) is calculated to check the agreement on the ranking of the results between two groups, and this method has been adopted in this book to compare the rankings of a subject between green and conventional construction projects. Without the consideration of normality or equal variance of data, the Spearman rank correlation coefficient (*S*) focuses on differences in rank orders of data rather than differences in means (Hwang et al., 2015a). The null hypothesis H_{000} is that there is a positive correlation between the means of the results for the two groups. The value of the coefficient ranges from -1 to 1, with 1 and -1 being the strongest positive and negative correlation, respectively. All hypothesis tests were conducted at a confidence level of 95% with a *P*-value of .05. If a *P*-value is less than .05, the null hypothesis was rejected.

3.4.1.5 Other Statistical Analysis Methods

Other statistical analysis methods, such as the Chi-square test and one-way analysis of variation (ANOVA), are used in some studies in this book. The Chi-square (χ^2) test is used to test whether there is a significant difference between the expected frequencies and the observed frequencies in one or more categories. For instance, to compare the productivity performance, the investigated projects were analyzed by considering the frequency of occurrence (hereinafter referred to FO) of productivity issues by using a Chi-square test. The Chi-square test was used to test whether FO is equally distributed between green and traditional construction projects. The null hypothesis H_0 for this test is that FO is statistically equally distributed between the green and traditional construction projects.

The one-way analysis of variance (ANOVA) test is a technique used to compare means of three or more samples. The ANOVA tests the null hypothesis that samples in two or more groups are drawn from populations with the same mean values. For instance, to test whether the project type has a significant effect on green cost premiums, the one-way ANOVA was performed. The null hypothesis H_0 is that there is no statistically significant difference in the cost premiums by project type, while the alternative hypothesis H_1 is that there is a statistically significant difference in the cost premiums by project type. Because the one-way ANOVA test does not show which specific building types differ significantly, the Tukey Post Hoc (TPH) test was performed to analyze the difference further.

3.4.2 Fuzzy Synthetic Evaluation Approach

This book addresses risks in green construction projects systematically and one ambitious research action is to develop a computational model to assess risks in such projects. As risk assessment is complex and ambiguous, qualitative linguistic terms are unavoidable. In addition, the perceptions of likelihood and impact of risk factors by respondents are typically subjective and uncertain. Thus, the fuzzy set theory is used in this book to develop the risk assessment model. Fuzzy set theory is a research approach that can deal with problems relating to ambiguous, subjective and imprecise judgments, and it can quantify the linguistic facet of available data and preferences for individual or group decision-making (Shan et al., 2015a). As an application of the fuzzy set theory, fuzzy synthetic evaluation was selected as the particular development approach as it aims to provide a synthetic evaluation of an object relative to an objective in a fuzzy decision environment with multiple criteria (Zhao et al., 2016). Fuzzy synthetic evaluation has been adopted widely in risk management studies. For instance, Mu et al. (2014) applied this method to assess the risk management capability of contractors in subway projects in Mainland China; Xu et al. (2010b) developed a fuzzy synthetic risk allocation model for public-private partnership projects in China; and Liu et al. (2013) used this method to analyze risks in ultra-deep scientific drilling projects. The advantage of the fuzzy synthetic evaluation approach mainly lies in dealing with complicated evaluation with multiple attributes and levels.

3.4.3 Risk Criticality Index

When examining the diverse risks in green retrofit and commercial and residential construction projects, this book solicited the respondents' perceptions on the likelihood of occurrence (LO) and magnitude of impact (MI) of each risk. Consequently, this book has adopted a risk criticality (RC) index to evaluate the criticality of each risk. RC has been widely recognized as a function of the LO and MI in many risk management studies (Zou et al., 2007; Wu and Feng, 2014; Ke et al., 2011), and the RC of a risk is computed as follows:

$$\mathrm{RC}_j^i = \mathrm{LO}_j^i \times \mathrm{MI}_j^i \tag{3.1}$$

$$\mathrm{RC}^i = \frac{1}{n}\sum_{j=1}^{n} \mathrm{RC}_j^i \tag{3.2}$$

where n = the number of the respondents; RC_j^i = the risk criticality of the risk i by respondent j; and RC^i = the risk criticality of the risk i. RC is on a full scale of 25, and thus a benchmark of 9 is established to identify the critical risks determined by the product of the median values of LO and MI rating scales. Under this benchmark, risks with RC values above 9 are considered as critical risks in green construction projects, and their distributions fall in Zone A, as shown in Fig. 3.2.

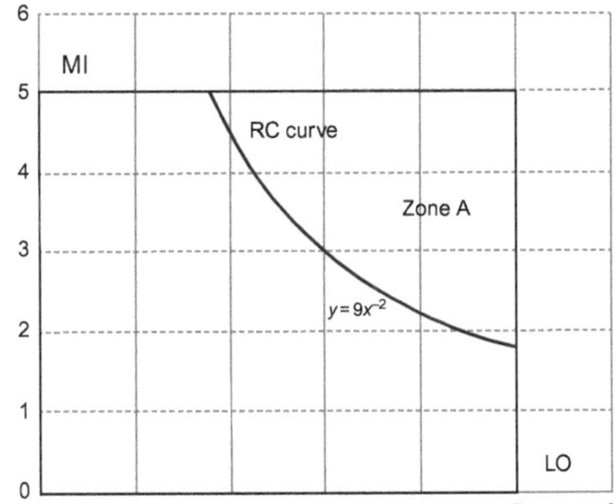

FIG. 3.2 RC curve for the critical risks in green construction projects.

3.5 SUMMARY

This chapter began by presenting the research design used in this book. Subsequently, the data collection methods adopted, namely, literature review, questionnaire survey and postsurvey interviews, were introduced in detail. Additionally, various data analysis approaches, such as parametric and nonparametric statistical tests, the fuzzy synthetic evaluation approach, and the risk criticality index, were fully explained.

Chapter 4

Green Building Rating Systems: Practices and Research Efforts

4.1 INTRODUCTION

Along with the rapid development of green buildings over the past two decades, a variety of green building rating systems (GBRSs) have emerged (Doan et al., 2017; Awadh, 2017). Generally, GBRS is a comprehensive framework developed by construction authorities, international organizations, or private consultancy companies with the intention of assessing and verifying the sustainability and greenness of buildings (Lee et al., 2013a; Toroghi et al., 2016). It includes a group of explicit performance thresholds that buildings must meet in order to be certified, as well as particular guidelines that can help project teams meet or exceed those performance thresholds (Mattoni et al., 2018; Wu et al., 2017). GBRS informs people how eco-friendly and environmentally sound a building is, depicts to what extent green components have been incorporated, and identifies the sustainable principles and practices that have been employed (Kubba, 2010; Pearce et al., 2012). GBRS has become increasingly vital in current green building development, as it can help building owners in the following respects: baselining (i.e., establishing an initial measurement against which to calibrate future performance), benchmarking (i.e., providing a basis for comparison with competitors), decision making (i.e., establishing a basis to choose among different solutions), and documentation (i.e., capture evidence to comply with sustainable rules and regulations) (Eisenstein et al., 2017; Park et al., 2017; Kubba, 2010, 2012).

Today, various GBRSs have been developed around the world to evaluate the sustainability of buildings (Li et al., 2017; Chen et al., 2015; Mattoni et al., 2018). Most of them are specifically tailored to suit the building industry of the country in which they are developed and to serve the specific needs of that country's buildings (Kubba, 2010, 2012). For example, in 2005, the Building and Construction Authority (BCA) of Singapore launched BCA Green Mark (GM) as a benchmarking scheme to differentiate buildings in Singapore in terms of the practices in environmental design and performance (Hwang et al., 2016, 2017b). In 2006, China's Ministry of Construction released Assessment Standard for Green Building (ASGB) as the local green building standard to create sustainability ratings for buildings across the country (Kubba, 2010,

Performance and Improvement of Green Construction Projects. https://doi.org/10.1016/B978-0-12-815483-0.00004-1

2012; Zhu et al., 2013; Ye et al., 2013, 2015). In the same year of 2006, the United Kingdom released Code for Sustainable Homes (CSH), an environmental assessment method designed to rate and certify the sustainability performance of new homes built in the United Kingdom (Department of Communities and Local Government, 2010; Zhang et al., 2017; Ganah et al., 2015). Derived from the GM of Singapore, Malaysia launched the Green Building Index (GBI) in 2009 as its first comprehensive rating system for evaluating the environmental design and performance of buildings (Green Building Index, 2013; Illankoon et al., 2017a,b). In 2009, Qatar launched the Global Sustainability Assessment System (GSAS), the first GBRS developed to rate and certify green buildings and infrastructure in the Middle East and North Africa region (Awadh, 2017; Rybkowski et al., 2017). All of these GBRSs have played a major role in raising awareness and in popularizing green buildings worldwide.

Given the wide application of GBRS over recent years, this chapter aims to conduct a comprehensive review of prevailing GBRSs used around the world (focusing on their timeline, particular versions, essential evaluation criteria, and the hierarchy of rating results). Existing GBRS studies are summarized and new directions for research in GBRS are proposed.

4.2 METHODS

Considerable research effort has been devoted to investigating the prevailing GBRSs adopted by the current sustainable construction industry. As a result, an increasing number of research articles relating to GBRS have been published (Zuo and Zhao, 2014; Olubunmi et al., 2016; Darko et al., 2017). Therefore, this section presents a systematic review of the GBRS papers to identify the prevailing GBRSs used by the current sustainable construction industry. Scopus is a powerful, widely recognized academic search engine used by researchers worldwide to track the latest research progress in various areas (Le et al., 2014; Rupp et al., 2015). Hence, Scopus was used to track down those GBRS papers published from 2008 to 2017. Four concise codes, including "TITLE: (green building) AND TITLE: (standard)," "TITLE: (green building) AND TITLE: (assessment)," and "TITLE: (green building) AND TITLE: (rating)," were searched in Scopus. Eventually, 25 papers directly relating to GBRS were obtained, from which 15 prevailing GBRSs were identified. These 15 GBRSs were Assessment Standard for Green Building (ASGB) from China, Building Research Establishment's Environmental Assessment Method (BREEAM) from the United Kingdom, Comprehensive Environmental Performance Assessment Scheme (CEPAS) from Hong Kong, Code for Sustainable Homes (CSH) from the United Kingdom, Comprehensive Assessment System for Building Environmental Efficiency (CASBEE) from Japan, Estidama Pearl Rating System (EPRS) from Abu Dhabi, Green Building Index (GBI) from Malaysia, Green Globes (GG) from North America, Green Mark (GM) from Singapore, Global Sustainability Assessment System (GSAS) from Qatar,

Green Star (GS) from Australia, Built Environmental Assessment Method (BEAM) from Hong Kong, Indian Green Building Council (IGBC) Rating system from India, Leadership in Energy and Environmental Design (LEED) from The United States, and the International SBTool (ISBT) developed by a non-profit organization named International Initiative for a Sustainable Built Environment. Table 4.1 presents the 15 GBRSs and their sources. From Table 4.1, it can be found that LEED, BREEAM, CASBEE, GS, ASGB, and BEAM were the GBRSs most frequently mentioned by the identified papers, suggesting they had attracted greater attention from academia than the rest of the GBRSs.

4.3 PREVAILING GBRSs: PRACTICE

4.3.1 Timeline of the Identified Prevailing GBRSs

Fig. 4.1 shows the timeline of the prevailing GBRSs. In 1990, the Building Research Establishment Ltd. in the UK developed BREEAM, the world's first sustainability rating scheme for the development of the built environment (BRE Global Ltd., 2017; Ferreira et al., 2014; Schweber, 2013). In 1994, the US Green Building Council released the famous LEED, one of the most popular and widely used GBRSs in the world so far (Zhang et al., 2017; Awadh, 2017). In 1996, the Hong Kong Green Building Council and the BEAM Society Limited developed BEAM, a local and voluntary GBRS to rate buildings in terms of sustainability and greenness (Chen et al., 2012, 2015). After entering the 21st century, an increasing number of GBRSs were launched worldwide. In 2000, the ECD Energy and Environment Canada developed GG, a GBRS that is used primarily in Canada and the United States (Green Building Initiative, 2017; Wu and Low, 2010). In 2001, the Ministry of Land, Infrastructure, Transport and Tourism of Japan developed CASBEE to evaluate and rate the environmental performance of buildings and the built environment in Japan (Chen et al., 2015; Wong and Abe, 2014). In 2002, the Buildings Department of the Hong Kong government devised CEPAS as the local official GBRS for buildings in Hong Kong (Awadh, 2017). In 2003, the Green Building Council Australia launched GS to regulate the design, construction, and operation of green buildings in Australia, and to rate buildings from the perspective of sustainability (Love et al., 2012; Zuo et al., 2016). In 2005, Singapore launched GM, a benchmarking scheme to drive Singapore's construction industry toward more environmentally friendly buildings (Wu and Low, 2010; Hwang et al., 2017b). In 2006, China issued ASGB to assess buildings in terms of sustainability and thereby to promote the development of green buildings across the country (Geng et al., 2012; Ye et al., 2013, 2015). In the same year, the United Kingdom released CSH as the national standard for the sustainable design and construction of new homes (Zhang et al., 2017; Ganah et al., 2015). In 2007, a nonprofit organization, the International Initiative for a Sustainable Built Environment, launched ISBT as a generic framework for rating the sustainable performance

TABLE 4.1 Identified GBRSs and References

No.	Reference	ASGB	BREEAM	BEAM	CEPAS	CSH	CASBEE	EPRS	GBI	GG	GM	GSAS	GS	IGBC	ISBT	LEED
1.	Haapio and Viitaniemi (2008)		√													√
2.	Retzlaff (2008)									√						√
3.	Retzlaff (2009)									√						√
4.	Ali and Al Nsairat (2009)		√				√									√
5.	Wu and Low (2010)									√	√					√
6.	Aye and Hes (2012)		√										√			√
7.	Love et al. (2012)												√			
8.	Geng et al. (2012)	√	√	√			√									√
9.	Ye et al. (2013)	√														
10.	Gou and Lau (2014)		√				√									√
11.	Kim and Osmond (2014)		√										√			√
12.	Chen et al. (2015)	√	√	√			√									√
13.	Miller et al. (2015)		√										√			√
14.	Yu et al. (2015)	√	√				√									√
15.	Ye et al. (2015)	√														

16.	Bozovic-Stamenovic et al. (2016)		√				√				√		√			√
17.	Wu et al. (2016)	√	√						√	√						√
18.	Zuo et al. (2016)		√										√			√
19.	Li et al. (2017)		√	√			√						√			√
20.	Zhang et al. (2017)	√				√										√
21.	Awadh (2017)		√	√	√			√		√		√			√	
22.	Huo et al. (2017)	√	√	√							√					√
23.	Illankoon et al. (2017a)		√	√			√		√		√		√	√		√
24.	Illankoon et al. (2017b)		√	√			√		√		√		√	√		√
25.	Rastogi et al. (2017)															√
Total		8	17	7	1	1	9	1	3	5	5	1	9	2	1	21

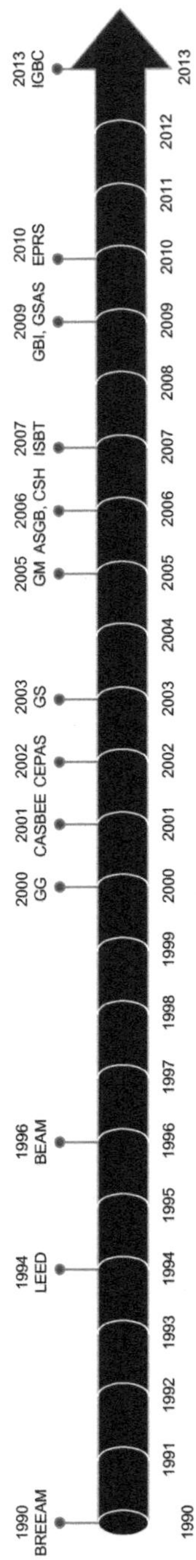

FIG. 4.1 Timeline of the prevailing GBRSs in effect.

of buildings and projects (Awadh, 2017). In 2009, Malaysia launched GBI as its industry-recognized green rating tool for buildings (Illankoon et al., 2017a,b). Qatar also launched GSAS in 2009, which was also the first GBRS developed for rating green buildings and infrastructure in the Middle East and North Africa region (Awadh, 2017). Subsequently, in 2010, the Abu Dhabi Urban Planning Council launched EPRS as a rating framework for sustainable design, construction, and operation of communities, buildings, and villas (Awadh, 2017). In 2013, the Indian Green Building Council launched IGBC to regulate, rate and certify the development of green buildings in India (Illankoon et al., 2017a,b). From this timeline, it can be noted that GBRS started to develop in the early 1990s, and that the United Kingdom and the United States are pioneers in developing GBRSs. After entering the 21st century, particularly in the first decade of the 21st century, a greater number (11) of GBRSs were developed and released by various countries around the world. Such a sharp increase can be ascribed to the increasing recognition of green buildings among construction industries and authorities worldwide, as well as to the good demonstration of the two pioneer GBRSs, namely BREEAM and LEED (Ferreira et al., 2014; Cole and Jose Valdebenito, 2013).

4.3.2 Particular Versions of Prevailing GBRSs

According to the review, the prevailing GBRSs offer a great number of particular versions in terms of project stages, project types, and building types. For example, CEPAS, GS, and LEED offer particular versions of GBRSs in terms of project stages, such as predesign (e.g., CEPAS for Pre-design Assessment), design (e.g., CEPAS for Design Assessment), construction (e.g., LEED for Building Design and Construction), and operation (e.g., LEED for Building Operations and Maintenance) (Buildings Department of Hong Kong, 2006; Love et al., 2012; Karakhan, 2016). BREEAM, CASBEE, GBI, GG, GM, and BEAM offer particular GBRS versions for particular project types, such as new construction (e.g., CASBEE for Buildings-New Construction and BEAM Plus New Buildings) and renovation (e.g., CASBEE for Buildings-Renovation and BEAM Plus Existing Buildings) (Schwartz and Raslan, 2013; Murakami et al., 2011; Green Building Index, 2013; Green Building Initiative, 2017; Hwang et al., 2016; Wong and Kuan, 2014). BREEAM, EPRS, GBI, GM, GSAS, GS, BEAM, IGBC, and LEED offer particular versions of GBRS in terms of building types, such as residential buildings (e.g., GM for Residential Buildings), commercial buildings (e.g., GSAS Commercial), industrial buildings (e.g., GBI Existing Industrial Building), infrastructure (e.g., BREEAM for Infrastructure Civil Engineering & Public Realm), healthcare (e.g., IGBC Green Healthcare), school (e.g., GSAS Schools), sports (e.g., GSAS sports), parks (e.g., GM for New and Existing Parks), and even restaurants (e.g., GM for Restaurants) (Seinre et al., 2014; Awadh, 2017; Green Building Index, 2013; Hwang et al., 2017c; Alhorr, 2017; Morris et al., 2018; Lee et al., 2007;

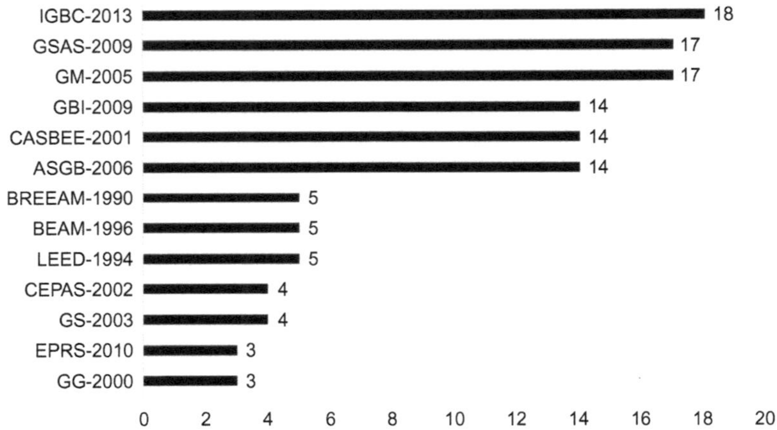

FIG. 4.2 The number of the particular versions for identified prevailing GBRSs.

Indian Green Building Council, 2014; Wu et al., 2017). Fig. 4.2 presents the number of the particular versions of the identified GBRSs. It can be seen that 13 out of 15 identified GBRSs have particular versions. Especially, ASGB, CASBEE, GBI, GM, GSAS, and IGBC have more than ten particular GBRS versions. Furthermore, by linking the identified GBRSs' particular version number and the years in which they came into effect, it can be found that there is a sharp increase in the number of particular versions after the year 2005, from 5 to 14. The reason for such a sharp increase is that, along with the rapid development of green buildings over recent years, the industry practitioners are calling for some particular GBRSs to help them address the implementation of green buildings in a more specialized manner (Haapio and Viitaniemi, 2008; Li et al., 2017; Zhang et al., 2017).

4.3.3 Essential Evaluation Criteria of Prevailing GBRSs

To provide a complete picture of the GBRS evaluation criteria, this section reviews and summarizes all the evaluation criteria extracted from the 15 prevailing GBRSs. Table 4.2 shows the 29 evaluation criteria that were identified in the review. Among the 29 criteria, "water," "material," "energy," "indoor environment," "site," "land and outdoor environment," and "innovation" were the most essential evaluation criteria due to their higher frequencies. Table 4.3 presents the weightings of the seven essential evaluation criteria, which were summarized from the manuals, instructions, and guidelines of the prevailing GBRSs. The average weighting of each essential evaluation criterion in different GBRSs was also calculated and presented in Table 4.3. It can be found that "energy" is the most important evaluation criterion due to the highest average weighting of 25%, following by "site" (20%), "indoor environment" (14.78%), "land and outdoor environment" (13.25%), "material" (10.98%), "water"

TABLE 4.2 Evaluation Criteria of the Prevailing GBRSs

No.	Evaluation Criterion	ASGB	BREEAM	CEPAS	CSH	CASBEE	EPRS	GBI	GG	GM	GSAS	GS	BEAM	IGBC	ISBT	LEED	Total
1.	Water	√	√	√	√	√	√	√	√	√	√	√	√	√	√	√	15
2.	Material	√	√	√	√	√	√	√	√	√	√	√	√	√	√	√	15
3.	Energy	√	√		√	√	√	√	√	√	√	√	√	√	√	√	14
4.	Indoor environment	√				√	√	√	√		√	√	√	√	√	√	11
5.	Site			√				√	√		√		√	√	√	√	8
6.	Land and outdoor environment	√	√	√	√	√	√					√					7
7.	Innovation	√	√				√	√				√		√		√	7
8.	Waste		√		√			√		√		√					5
9.	Construction project Management	√	√						√			√					4
10.	Operation management	√			√						√						3
11.	Health and wellbeing		√		√					√							3
12.	Transport		√									√				√	3

Continued

TABLE 4.2 Evaluation Criteria of the Prevailing GBRSs—cont'd

No.	Evaluation Criterion	ASGB	BREEAM	CEPAS	CSH	CASBEE	EPRS	GBI	GG	GM	GSAS	GS	BEAM	IGBC	ISBT	LEED	Total
13.	Integrative process						√									√	2
14.	Quality of service					√									√		2
15.	Loadings			√											√		2
16.	Social, cultural and perceptual Aspects										√				√		2
17.	Cost and economic aspects										√				√		2
18.	Pollution		√		√												2
19.	Surface water run-off				√												1
20.	Urban connectivity										√						1
21.	Sustainable architecture and design													√			1
22.	Building amenities			√													1

23.	Neighborhood amenities			√													1
24.	Neighborhood impacts			√													1
25.	Off-site environment					√											1
26.	Climatic responsive design									√							1
27.	Advanced green efforts									√							1
28.	LEED accredited professional															√	1
29.	Regional priority															√	1

TABLE 4.3 Weightings of the Essential Evaluation Criteria of the Prevailing GBRSs

No.	GBRS	Energy (%)	Site (%)	Indoor Environment (%)	Land and Outdoor Environment (%)	Material (%)	Water (%)	Innovation (%)
1.	ASGB	19	–	14	17	14	16	9
2.	BREEAM	19.97	–	–	7.69	12.5	6.73	10
3.	CEPAS	–	23.2	–	11.6	6.13	2.04	–
4.	CSH	36.4	–	–	12	7.2	9	–
5.	CASBEE	20	–	20	15	12	3	–
6.	EPRS	24	–	8.53	19.47	16	24	2
7.	GBI	23	33	12	–	8	12	8
8.	GG	39	12	16	–	13	11	–
9.	GM	18	–	–	–	9.38	9.38	–
10.	GSAS	24	13	16	–	10	16	–
11.	GS	25	–	16	10	14	10	9
12.	BEAM	35	25	20	–	8	12	–
13.	IGBC	28	14	12	–	16	18	7
14.	ISBT	5	29	14	–	2.5	2.5	–
15.	LEED	28	12	14	–	16	9	4
Average weighting for each criterion		25	20	14.78	13.25	10.98	10.71	7
Standard deviation		9	8.4	3.4	4.1	4.1	6.1	2.9
Average weighting for all the criteria		14.53						

Note: "–" represents "not applicable."

(10.71%), and "innovation" (7%). Additionally, to quantify the weighting variation of each essential evaluation criterion, the standard deviations were calculated and are presented in Table 4.3. It can be found that the weighting of "innovation" shows the slightest variation due to the lowest standard deviation of 2.9%, while the weightings of "site" and "energy" show greater variations due to higher standard deviations, namely 8.4% and 9%, respectively. Due to space limitations, the top seven essential evaluation criteria are examined and discussed below.

4.3.3.1 Energy

According to the review, the criterion of "energy" mainly encompasses the reduction of energy use (particularly peak load reduction), energy monitoring and reporting, energy efficient appliances, equipment and systems, and the use of renewable energy (Abu Dhabi Urban Planning Council, 2010; Institute for Building Environment and Energy Conservation, 2014; BRE Global Ltd., 2017; BEAM Society Limited, 2012). According to Table 4.3, "energy" is the most important evaluation criterion because it receives the highest average weighting of 25%. Also, it can be found from Table 4.3 that the average weighting of "energy" is much higher than the average weighting for all the identified criteria (i.e., 14.53%), and that 13 out of 14 GBRSs that include the "energy" criterion have higher energy weightings than 14.53%. These results indicate that "energy" is an important evaluation criterion and used widely by different GBRSs in assessing green buildings. Moreover, it is noteworthy that, among the 14 GBRSs having the evaluation criterion of "energy," 10 GBRSs allocated the highest weighting to "energy" compared to the rest of the evaluation criteria. This further indicates that "energy" is the foremost evaluation criterion for assessing and certifying green buildings. A similar finding was also obtained by Illankoon et al. (2017b) who stated that the "energy" criterion had the highest consideration in the majority of existing GBRSs. Nevertheless, there is an outlier, which is ISBT. In ISBT, "energy" has a low weighting of 5%. This might be because there are only two indicators established in ISBT to assess building performance in "energy." These two indicators are "total life cycle nonrenewable energy" and "peak demand for facility operations" (Larsson, 2016).

4.3.3.2 Site

The criterion of "site" mainly includes site location, planning, design, management, the off-site services available, site assessment, and site regeneration and development (US Green Building Council, 2017; Larsson, 2016; Green Building Initiative, 2017; BEAM Society Limited, 2012; Indian Green Building Council, 2014). According to Table 4.3, "site" is mentioned by eight identified GBRSs including CEPAS, GBI, GG, GSAS, BEAM, IGBC, ISBT, and LEED, and it is the second most important evaluation criterion with the second highest average weighting of 20%. Also, it can be found that the average weighting for "site" (20%) is relatively higher than the average weighting for all the identified criteria (14.53%), indicating that "site" is also one of the foremost criteria recognized by the industry and authorities in assessing and certifying green buildings. Particularly, in GBI and ISBT, it can be found that "site" was allocated high weightings of around 30%.

4.3.3.3 Indoor Environment

Receiving the third highest average weighting of 14.78%, "indoor environment" becomes the third most important evaluation criterion of prevailing GBRSs. This criterion was mentioned by 11 GBRSs including ASGB, CASBEE, EPRS, GBI, GG, GSAS, GS, BEAM, IGBC, ISBT, and LEED. This criterion covers a wide range of indicators relating to the indoor environment, such as the sound environment, thermal comfort, lighting and illumination, healthy ventilation delivery, air quality, hygiene, occupant well-being facilities, and visual and acoustic comfort (Institute for Building Environment and Energy Conservation, 2014; Abu Dhabi Urban Planning Council, 2010; Green Building Index, 2013; Green Building Council, 2016; BEAM Society Limited, 2012; Indian Green Building Council, 2014; Larsson, 2016; US Green Building Council, 2017). Additionally, it is noteworthy that "indoor environment" has a low standard deviation of 2.9% compared to other evaluation criteria, indicating that "indoor environment" was allocated similar weightings among different GBRSs.

4.3.3.4 Land and Outdoor Environment

"Land and outdoor environment" can be regarded as the fourth most important evaluation criterion as it receives the fourth highest average weighting, namely, 13.25%. This criterion is included in seven GBRSs including ASGB, BREEAM, CEPAS, CSH, CASBEE, EPRS, and GS. Major indicators under this criterion concern the reuse of land, remediation of contaminated land, ecological enhancement, long term impact on biodiversity, habitat creation and restoration, improved outdoor thermal comfort, accessible community facilities, active urban environments, private outdoor space, and public transport (Institute for Building Environment and Energy Conservation, 2014; Green Building Council, 2016; Department of Communities and Local Government, 2010; BRE Global Ltd., 2017; Abu Dhabi Urban Planning Council, 2010). Furthermore, Table 4.3 shows that the standard deviation of the weightings for "land and outdoor environment" in different GBRSs, at 4.1%, is relatively low compared to other evaluation criteria, suggesting that "land and outdoor environment" receives consistent weighting allocations among different GBRSs.

4.3.3.5 Material

The evaluation criterion of "material" mainly assesses buildings in terms of material sourcing, the efficient use of material, the use of recycled and certified material, and the environmental impact of the material (US Green Building Council, 2017; Institute for Building Environment and Energy Conservation, 2014; Indian Green Building Council, 2014; Green Building Index, 2013; Department of Communities and Local Government, 2010; Buildings Department of Hong Kong, 2006; BRE Global Ltd., 2017; BEAM Society Limited, 2012; Abu Dhabi Urban Planning Council, 2010). From Table 4.3, it can be found that, although "material" was mentioned by all the identified GBRSs, it received just the fifth highest average weighting (10.98%) and thus can only be the fifth most important evaluation criterion. Among the 15 prevailing GBRSs, only IGBC has a "material" weighting (16%) that is higher than the average weighting for all the identified criteria (14.53%). Particularly, in ISBT, the weighting allocated to "material" is merely 2.5%. It refers to one indicator only, which checks "the use of material" in the assessed building (Larsson, 2016).

4.3.3.6 Water

"Water" is the sixth most important evaluation criterion with an average weighting of 10.71%. This evaluation criterion was mentioned by all the 15 identified GBRSs. According to the review, this criterion mainly encompasses water quality, water saving performance, water cycling, the use of water efficient appliances, and alternative water sources (Building and Construction Authority, 2017; US Green Building Council, 2017; Department of Communities and Local Government, 2010; BEAM Society Limited, 2012). From Table 4.3, it can be found that the weightings of "water" in different GBRSs range from 2.04% to 24%. The weightings in CEPAS, ISBT, and CASBEE are all very low at less than 3%. This is because in CEPAS, ISBT, and CASBEE the criterion of "water" was included under the higher level criterion of "resources" and was only gauged by one or two indicators (Buildings Department of Hong Kong, 2006; Larsson, 2016; Institute for Building Environment and Energy Conservation, 2014). By contrast, the weighting of "water" in EPRS reaches up to 24% (Abu Dhabi Urban Planning Council, 2010), which is almost 12 times that of CEPAS, ISBT, and CASBEE. This might be because EPRS originates in Abu Dhabi, a city located in an arid land where fresh water is scarce. Thus local authorities in Abu Dhabi place much importance on water efficiency (Assaf and Nour, 2015).

4.3.3.7 Innovation

The criterion of "innovation" is mentioned by seven prevailing GBRSs including ASGB, BREEAM, EPRS, GBI, GS, IGBC, and LEED. Primarily, this criterion was included in GBRSs with the intention of encouraging, recognizing, and rewarding innovative technologies, designs, processes and strategies that can promote sustainability in the built environment and help transform the current construction market and industry toward greenness (Indian Green Building Council, 2014; Green Building Index, 2013; Green Building Council, 2016; BRE Global Ltd., 2017; Abu Dhabi Urban Planning Council, 2010). It is noteworthy that "innovation" is a new evaluation criterion appearing only in recent years (Illankoon et al., 2017b). From Table 4.3, it can be found that "innovation" has a relatively low average weighting of 7% and ranks as the seventh most important evaluation criterion for GBRSs. This is because the criterion of "innovation" normally contains a few open and flexible questions on the innovation assessment, and thus can only have limited weightings (US Green Building Council, 2017; Green Building Index, 2013; BRE Global Ltd., 2017).

4.3.4 Hierarchy of the Rating Results of Prevailing GBRSs

The hierarchy of the rating results provided by the prevailing GBRSs are summarized in Table 4.4. Seven prevailing GBRSs, namely CEPAS, GBI, GM, BEAM, IGBC, ISBT, and LEED, use a four-level hierarchy rating system (certified/bronze, silver, gold, and platinum) to rate and certify green buildings. Four prevailing GBRSs, namely BREEAM, CASBEE, EPRS, and GG, use a five-level hierarchy rating system (pass, good, very good, excellent, and outstanding) to rate and certify green buildings. Three prevailing GBRSs, namely CSH, GSAS, and GS, use a six-level hierarchy rating system (one-star, two-star, three-star, four-star, five-star, and six-star) to rate and certify green buildings. The only GBRS left, namely the ASGB from China, uses a three-level hierarchy rating system (one-star, two-star, and three-star). This brief summary shows that the majority of the identified GBRSs are using hierarchy rating systems that contain at least four levels to rate green buildings. The rationale behind this is that using a hierarchy system with a large number of levels can encourage building owners to keep upgrading their existing buildings or constructing their new buildings to a higher level of sustainability, and thereby achieving an

TABLE 4.4 The Hierarchy of the Rating Results of Prevailing GBRSs

Green Building Rating System	The Hierarchy of the Rating Results	Quantity
ASGB	3	1
CEPAS, GBI, GM, BEAM, IGBC, ISBT, LEED	4	7
BREEAM, CASBEE, EPRS, GG	5	4
CSH, GSAS, GS	6	3

overall improvement on sustainability in the built environment. Let us now turn to the scoring of the prevailing GBRSs. Basically, the identified GBRSs rate buildings based on the overall scores the buildings achieve from the assessment. However, several identified GBRSs stipulate some mandatory options for buildings in addition to referring to the overall scores. For example, ASGB established 76 options to certify green residential buildings. These options include 27 mandatory options, 40 regular options, and nine premium options, and any residential building to be certified by the ASGB must complete the 27 mandatory options first (Geng et al., 2012). EPRS from Abu Dhabi, GM from Singapore, and CEPAS and BEAM from Hong Kong have stipulated similar requirements (Abu Dhabi Urban Planning Council, 2010; Building and Construction Authority, 2017; BEAM Society Limited, 2012; Buildings Department of Hong Kong, 2006).

4.4 PREVAILING GBRS: RESEARCH EFFORTS

4.4.1 Comprehensive Reviews of GBRSs Used Worldwide

Several studies have conducted comprehensive reviews of various GBRSs adopted by the current sustainable construction community. Haapio and Viitaniemi (2008) carried out a comprehensive review of the different GBRSs used worldwide and compared them in terms of the type of assessed buildings, the user of the tools, the phases of the building life cycle, the databases of the tools, and the result forms of the tool. In spite of achieving some valuable findings, Haapio and Viitaniemi (2008) admitted that it was difficult to compare different GBRSs as they were designed to focus on different phases of the building life cycle and to assess different types of buildings. Illankoon et al. (2017b) reviewed eight prevailing GBRSs, including LEED, BREEAM, GS, GM, GBI, BEAM, IGBC, and CASBEE, and identified seven key evaluation criteria used by these GBRSs, namely, "site," "energy," "water," "indoor environment quality," "material," "waste and pollution," and "management." Illankoon et al. (2017b) compared these key evaluation criteria and found "energy" was the most widely considered evaluation criterion, following by "indoor environment quality" and "water." Li et al. (2017) conducted a systematic review of the existing literature on GBRSs. In their review, Li et al. (2017) identified the main contributing authors (e.g., Wai-Ling Lee from the Hong Kong Polytechnic University, Haapio Appu from Helsinki University of Technology, and Cole Raymond from the University of British Columbia), the represented countries and regions (e.g., China, United Kingdom, and Hong Kong), and the popular peer-reviewed journals (e.g., Energy and Buildings, Building and Environment, Environmental Impact Assessment Review, and Renewable and Sustainable Energy Reviews) in the area of GBRS research. In addition, Li et al. (2017) compared different GBRSs with regard to their developers, certification levels, and evaluation criteria.

4.4.2 Reviews and Discussions of Regional GBRSs

Some studies have reviewed and discussed the GBRSs applied in particular countries, including China, Australia, Singapore, and the United States. Geng et al. (2012) introduced the history of ASGB, the GBRS from China, and compared it with other prevailing GBRSs like BREEAM, LEED, BEAM, and CASBEE. Ye et al. (2013, 2015) also made critical reviews of ASGB. Particularly, Ye et al. (2015) identified the major problems in applying ASGB in China, such as "users may be confused by evaluation standards," "contents in most standards largely overlap," "linkage between evaluation standards and other standards are not well," and "some mandatory provisions in local standards are unnecessary." Furthermore, Ye et al. (2015) proposed some helpful measures to solve these problems. Kim and Osmond (2014) carried out a cross-sectional survey of hospital patients, staff, visitors, and members of the public, to solicit their perceptions of the weightings of different evaluation criteria in GS-Healthcare Building, a particular GBRS established for healthcare buildings in Australia. Kim and Osmond (2014) found that "indoor environmental quality" was acknowledged by the survey respondents as the most significant evaluation criterion to certify green healthcare buildings. Issues relating to "transport and accessibility" were perceived as more significant by the survey respondents than the current weighting in GS–Healthcare Building. Environmental impacts relating to "emissions to air and water" were also identified as more significant by the survey respondents than the current weighting. Miller et al. (2015) conducted a survey with 53 professionals from the construction industry of Australia to solicit their perceptions of GS. The survey respondents opined that there should be an increase in the embodied energy consideration of structures in GS because the embodied energy made a major contribution to the life cycle consumption. Zuo et al. (2016) investigated the challenges and barriers in obtaining GS-Office Building ratings in Australia. Zuo et al. (2016) collected and analyzed the scoring sheets from the ratings of 264 certified office buildings and found that the credits related to the attributes of innovation, ecology and energy were comparatively difficult to achieve. Furthermore, Zuo et al. (2016) found that a large number of projects did not apply for the specific credits of refrigerant global warming potential, re-use of façade, topsoil and fill removal from site, and individual comfort control. Bozovic-Stamenovic et al. (2016) carried out a survey to investigate the public perception of GM, the GBRS applied in Singapore. Bozovic-Stamenovic et al. (2016) found that the public was satisfied with GM's biophilic criteria, such as greenery and daylight, and hoped the current GM framework could be expanded to include some features and strategies that addressed social and psychological well-being. Rastogi et al. (2017) introduced the various versions of LEED and presented the evolution of the sustainability parameters and credit scoring schemes of LEED over the past decade.

4.4.3 Comparison Between Different GBRSs on Particular Aspects

A large number of studies have concentrated on the comparison between different GBRSs on some particular aspects, such as project management, energy efficiency, waste management, and site management. For example, Wu and Low (2010) compared LEED, GG, and GM in terms of project management. Based on their comparison, they found that about 20% of the credits were allocated to the items related to project management and that there was a trend toward project management processes that can offer a low-cost, low-barrier avenue. Chen et al. (2015) investigated the evaluation criteria relating to passive design in five GBRSs, including BREEAM, LEED, CASBEE, BEAM, and ASGB, which were primarily established to improve the energy efficiency of green buildings. Chen et al. (2015) identified some deficiencies in existing criteria and suggested some improvements, such as "the passive design strategies should be more accurately defined and more comprehensive in scope in GBRS" and "the weighting allocated to each design parameter should be based on the relative influence of that parameter over environmental quality and/or energy use." Wu et al. (2016) compared five selected GBRSs (LEED, BREEAM, GG, ASGB, and GBI) in terms of waste management. They found that, of the five GBRSs compared, the highest relative significance index of waste management was given to ASGB, while the lowest relative significance index went to GBI. Also, Wu et al. (2016) checked the five GBRSs in terms of the common 3R principles, namely, reduce, reuse, and recycle, and found that ASGB and GG focused more on the reduce principle, while the other three systems concentrated more on the reuse and recycle principles.

Illankoon et al. (2017a) analyzed the effectiveness of eight prevailing GBRSs (LEED, BREEAM, GS, GM, GBI, BEAM, IGBC, and CASBEE) in evaluating environmental, economic, and social sustainability in buildings. They found that the main focus in prevailing GBRSs were environmental and social sustainability and that there was a lack of attention given to economic sustainability. Awadh (2017) conducted a similar comparison between LEED, BREEAM, EPRS, and GSAS in terms of environmental, social, and economic sustainability. Awadh (2017) came to a different conclusion from Illankoon et al. (2017a). Awadh (2017) found that all four GBRSs were focused on environmental sustainability while giving social sustainability the least importance. Huo et al. (2017) conducted a comparative analysis between BREEAM, LEED, BEAM, GM, and ASGB in terms of "site planning and design" and found that BEAM allocated the highest importance while GM allocated the lowest. Also, Huo et al. (2017) found that each of them emphasized different aspects of "site planning and design" in green buildings and BEAM involved the most "site planning and design" related indicators. In addition, Zhang et al. (2017) compared the ASGB from China, the CSH from the United Kingdom, and the LEED from the United States in terms of five aspects, including energy saving, water saving, material saving, site selection, and the outdoor and indoor environmental quality, and found that there was a large overlap in the criteria of indoor environmental quality among these three GBRSs.

4.4.4 Development of Particular GBRSs Tailored to Certain Purposes

Another focus of the existing GBRS research is on the development of particular GBRSs suited to some particular purpose. For example, on the basis of ASGB, Yu et al. (2015) developed a particular rating method to assess green store buildings in China. After studying the international GBRSs, such as LEED, CASBEE, BREEAM, and ISBT, Ali and Al Nsairat (2009) developed a green building rating system for rating the residential units in Jordan. Retzlaff (2009) provided planners with an introduction to the concept of GBRS and explored the major themes in the literature, including scope, weighting, subjectivity, rigor, adaptation, and life cycle analysis. Furthermore, referring to a series of GBRSs used by various countries around the world, Retzlaff (2008) proposed a GBRS framework from the perspective of planners. Moreover, Retzlaff (2008) highlighted that planners should treat GBRS seriously to make sure that their planning meets the specific sustainability goals established.

4.5 FUTURE RESEARCH DIRECTIONS

4.5.1 The Investigation of the Barriers and Drivers for Using GBRS

According to the review, the existing research efforts in the GBRS area mainly focus on the review of different GBRSs in the world, as well as the comparison between different GBRSs in some particular aspects. So far, however, limited research has been done to investigate the implementations of GBRSs, especially regarding the barriers and drivers for the use of GBRS. Currently, GBRSs are mainly voluntary in the market, which means it is not mandatory for building owners to obtain a GBRS rating for their buildings (Gowri, 2004; Kibert, 2016; Mao et al., 2009). Nevertheless, it would be better if most building owners could accept GBRS ratings as it really can help create a built environment that is more sustainable and environmentally friendly. Therefore, it is urgently necessary to investigate the barriers that hinder building owners from using GBRS ratings, as well as to explore valid drivers that can help building owners accept GBRS ratings.

4.5.2 The Examination of Project Stakeholders' Perceptions of GBRSs

In spite of the vast research regarding GBRSs, little has been done to examine the project stakeholders' perceptions of GBRSs. Without doubt, a careful examination of project stakeholders' perceptions of GBRSs can help reveal their strengths and deficiencies, and subsequent amendments based on these perceptions can help improve GBRSs accordingly. This can also help GBRS achieve better efficacy in terms of green building rating. Therefore, it is imperative to examine the project stakeholders' perceptions of GBRSs. As there have been numerous GBRSs developed by a great number of countries worldwide since the first launch of GBRS (i.e., BREEAM) in 1990, it is necessary to examine project stakeholders' perception of each developed GBRS within the particular context of its home country. Furthermore, as the majority of the prevailing GBRSs have a wide scope, including building design, construction, operation, and maintenance, it is strongly recommended that the perceptions from all the project stakeholders, including the owner, designer, contractor, supplier, and the building occupants, be examined comprehensively.

4.5.3 The Development of a Corporate Sustainability Rating System

In addition to rating projects and building components, another direction for future research is rating companies in terms of sustainability, focusing, in particular, on their capabilities in implementing green building projects. Such a company-based sustainability rating would be of particular use to clients, especially to those who award green building construction project contracts. This is because the results of such a corporate sustainability rating can easily reveal a company's competency and potential to be a sustainability leader. To realize such a corporate sustainability rating, a comprehensive rating framework should be developed. It is suggested that this framework would not only assess a company in terms of a broad set of metrics that reflect its professionals, equipment, financial capability, and risk management capability in terms of implementing green building construction projects, but also look at its past history in successfully implementing green building construction projects.

4.6 SUMMARY

GBRSs, which play a vital role in promoting the development of green buildings, have drawn considerable attention from construction authorities, industry, and academia worldwide in recent years. The aims of this chapter were to carry out a comprehensive review of prevailing GBRSs adopted around the world, to identify the major research efforts described in existing GBRS-related studies, and to reveal the directions for the future of GBRS research. Based on a systematic search of GBRS-related studies published during 2008–17, a total of 15 prevailing GBRSs, including ASGB from China, BREEAM from the United Kingdom, CEPAS from Hong Kong, CSH from the United Kingdom, CASBEE from Japan, EPRS from Abu Dhabi, GBI from Malaysia, GG from Canada, GM from Singapore, GSAS from Qatar, GS from Australia, BEAM from

Hong Kong, IGBC from India, LEED from the United States, and ISBT, developed by a nonprofit organization named International Initiative for a Sustainable Built Environment, were identified. The summary of the timeline showed that BREEAM, LEED, and BEAM were the early GBRSs launched in the 1990s, while the rest were primarily released in the first decade of the 21st century. The review also showed that the prevailing GBRSs offered a wide variety of particular versions in terms of project stages, project types, and building types. Furthermore, the review identified seven essential evaluation criteria that are commonly used by different prevailing GBRSs, and found that the most important evaluation criterion was "energy," following by "site," "indoor environment," "land and outdoor environment," "material," "water," and "innovation." This chapter also revealed that the majority of the identified GBRSs use a four-level hierarchy rating system or above to rate and certify green buildings. Additionally, this chapter identified four major themes within the existing GBRS research, which are "comprehensive review of GBRSs used worldwide," "reviews and discussions of regional GBRSs," "comparison between different GBRSs on particular aspects," and "the development of particular GBRSs tailored to certain purposes." Furthermore, this chapter also pointed out three directions for future GBRS research, including "the investigation of the barriers and drivers for using GBRS," "the examination of project stakeholders' perceptions of GBRSs," and "the development of a corporate sustainability rating system."

Despite the limitation that this chapter could not review all the GBRSs used in the countries around the world, the findings are still valuable. This chapter contributes to the body of knowledge by reviewing the prevailing GBRSs in terms of timeline, particular versions, essential evaluation criteria, and the hierarchy of rating results, by summarizing the major research efforts described in the existing GBRS studies, and by positing directions for the future of GBRS research. Furthermore, the findings from this chapter can provide industry practitioners with a global view of GBRS and thereby deepen their understanding of this topic.

Chapter 5

Obstacles and Solutions for Green Construction Projects

5.1 INTRODUCTION

Green buildings are constructed based on the principles of sustainable construction, which address the ecological, social and economic issues of a building in the context of its community. These buildings are designed and built to use less energy and resources than conventional buildings and aim to minimize their impacts on the environment (Environmental Protection Agency, 2016). Recently, along with the increasing prominence of green construction, a growing emphasis has been placed on project management approaches. Therefore, the specific objectives of this chapter are (1) to investigate the common obstacles that practitioners encounter during the planning and construction of green building projects and (2) to propose feasible solutions to improve the current green construction project management approaches, thus leading to a more intensive implementation of green construction projects worldwide.

Current literature shows that practitioners of green construction projects may encounter obstacles, such as high cost premiums, unequal distribution of benefits, lack of green product information, complex legislation, and lack of awareness.

5.1.1 High Cost Premium

Normally, it costs more to construct green buildings as compared to conventional buildings (Yudelson, 2010a) because green materials cost significantly more than ordinary materials (Kibert, 2008). For example, compressed wheat board, which is a green substitute for plywood, costs about ten times more than ordinary plywood. Other factors contributing to the high cost premium of green buildings are the costs incurred in the search for green alternatives and the certification of buildings (Yudelson, 2010a). Shi et al. (2013) discovered that using green techniques on site results in additional cost, which discouraged the adoption of such methods. Therefore, when managing green construction projects, it is difficult to keep within the project budget.

Performance and Improvement of Green Construction Projects. https://doi.org/10.1016/B978-0-12-815483-0.00005-3

5.1.2 Unequal Distribution of Benefits

It is difficult to convince a developer to build green when there is an unequal distribution of benefits between the builder and tenants (Yudelson, 2010a). Developers have to pay the high cost premium for green buildings, while the tenants accrue most of the benefits, such as better indoor environmental quality and cost savings in energy and water. In addition, the extra cost incurred by the green building cannot be readily passed on to the tenants (Architecture Week, 2001).

5.1.3 Lack of Green Product Information

Currently, there is still inadequate information regarding green products and sustainable building systems that can be implemented in a green building (Architecture Week, 2001). Hence, developers are being forced to engage consultants, who are specialized in green products and building systems, for a fee. Without adequate information, the developer can also risk losing green certification or incur additional cost to correct products or systems that do not meet specified green standards.

5.1.4 Complex Legislation

Lately, green building codes and regulations have been getting more complicated, causing difficulties to developers when evaluating the cost involved in complying with such codes (Architecture Week, 2001). Often, developers fail to see convincing benefits in green building, and thus do not feel incentivized to go "green."

5.1.5 Lack of Awareness

The traditional perception of how a building should be constructed still prevails and many developers resist building green due to the perceived risks (Kibert, 2008). Shi et al. (2013) found that there is a lack of awareness of, and readiness to adopt, environmental auditing. Another study proves the lack of awareness about green building by showing that contractors manage and use materials without giving much thought about sustainability and the surrounding environment (Chew, unpublished undergraduate dissertation). Also, there is a lack of awareness among the public regarding the benefits of green buildings due to insufficient research, especially on issues such as the effects of indoor air quality of green buildings on productivity and health (Kibert, 2008).

5.2 METHOD AND DATA PRESENTATION

To achieve the research objectives of this chapter, literature reviews on green and sustainable building construction, including the management approach of such projects in Singapore and overseas, were carried out first. The aim of the reviews is to explore the differences between conventional and green construction projects and to identify the obstacles encountered in green building construction project management overseas. Another objective of the reviews is to discover innovative management skills applied in these projects that can be implemented in Singapore's context. The examination and analysis of past work related to green construction also provides a better understanding of green and sustainable building construction and the project management processes involved.

As in Chapter 3, questionnaire survey data play a key role in this chapter. The aim of the survey was to capture possible obstacles encountered in the management of local green building construction projects in Singapore that are not reflected in the literature review. The survey also provided a comprehensive and accurate outlook of management culture adopted by the local green construction industry. The obstacles of green building project management and their respective solutions provided in the survey questionnaire were formed based on the findings derived from the literature reviews.

The developed questionnaire consisted of four sections. The first section asked about the details of the respondents. The second section carried questions regarding the profile of green construction projects undertaken by the respondents. The third section asked the respondents about the obstacles encountered during management of green construction and the fourth section requires respondents to choose solutions to overcome those obstacles.

The questionnaires were sent, via email, to a population size of 101 managers and professionals listed under the BCA's Certified Green GMM and GMP Scheme. They were chosen as the target population because they have a strong foundation in, and a deep knowledge of, green construction, and have the professional capability to advise on the design of environmentally friendly buildings (BCA, 2009). It is believed that with their experiences in planning, tender and procurement, construction, and commissioning of green building projects, they can provide comprehensive accounts of the obstacles faced by the management of such construction projects.

In addition to the questionnaire survey, postsurvey interviews with ten GMPs and GMMs, who have managed projects that had received either Green Mark Platinum or Gold Plus award, and who have more than 3 years of experience in the field of green construction, were conducted. The goals of the interviews were to: (1) gain deeper insight and understanding of the obstacles listed in the survey questionnaire; and (2) get experts' opinions on the survey results. Through supplementary knowledge and valuable opinions from managers and professionals, quality findings and discussions were produced.

Out of the questionnaires disseminated, 31 completed copies were received. The survey results were analyzed using a simple tabulation of percentages and are presented in Table 5.1. The respondents were from 19 consultancies (61%) and 12 project management companies (39%). All of the respondents had at least 2 years of experience on green construction projects, and the majority

TABLE 5.1 Obstacles Encountered During Green Building Project Management

No.	Obstacles	Frequency	Percentage (%)
1	High cost premium of green building project	31	100.0
2	Lack of communication and interest among project team members	26	83.9
3	Lack of required expertise in green building	0	0.0
4	Lack of knowledge regarding green building principles	0	0.0
5	Lack of management and time to implement green construction practices	8	25.8
6	Lack of expressed interest from client and market demand	16	51.6
7	Lack of expressed interest from other project team members	12	38.7
8	Resistance to change from conventional to green practices by company's employees	11	35.5
9	Lack of government support (incentives) for sustainable construction	0	0.0
10	Green building practices are costly to implement	24	77.4
11	Lack of information regarding green products and building systems	9	0.0
12	Lack of credible research on the benefits of green buildings	21	67.7
13	Complex codes and regulations on green building and sustainable construction	7	22.6

of them (19 out of 31 respondents; 61%) had 3–4 years of experience. Also, there were five respondents (16%) who had more than 4 years of experience in the area.

5.3 ANALYSIS RESULTS AND DISCUSSIONS

This section elaborates results of the survey provided by the 31 industry experts.

5.3.1 Delivery Systems Adopted for Green Construction Projects

This section investigates the delivery systems currently adopted for green construction projects. The survey results indicate that 68% of the surveyed green construction projects (21 out of 31 projects) adopted the conventional design-bid-build delivery system, while no project chose construction manager at-risk or project manager at-risk as their delivery systems. Comparatively, project manager for a fee and design-build delivery systems were adopted by 10% (three projects) and 22% (seven projects), respectively. This result was contrary to what Kibert (2008) had mentioned in his book, namely, more use of design-build than design-bid-build. Apparently, most of the green building construction projects performed in Singapore adopted the conventional design-bid-build delivery system instead of design-build. One possible reason for this is that Singapore's construction industry lacks a comprehensive project delivery system that is well-suited for green building construction projects. Hence, project managers and contractors manage green projects using the conventional method, which is something familiar and stable to them.

In addition, all the respondents indicated that for the projects they undertook, other parties, such as Ecologically Sustainable Development (ESD), Green Mark, Total Building Performance (TBP), Acoustic, and Security and Lighting Consultants, were engaged to assist in the construction of the green building. This finding coincides with the article "Barrier to Building Green from Architecture Week (2001)," which states that a lack of green product information compelled developers to engage external consultants who specialized in green products and building systems. Some respondents stated that hiring ESD and green consultants leads to cost-related problems. Their clients often perceive the cost of engaging these consultants to be included in conventional consultation fees. However, in most projects, the cost involved for such consultancy is an additional amount on top of conventional consultation services, inflating the cost premium required to design and construct a green building.

It is also of interest that all respondents stated that the decision to incorporate green features into the project was made at the very beginning during the planning and feasibility studies stage of their projects. This suggests that the respondents' clients or developers already had the vision of building green. Such a decision makes good economic sense because incorporation of green features at later project stages would be expensive as a result of multiple changes.

Similar results were obtained from the postsurvey interviews. The interviewees indicated that, although all conventional construction projects adopt least-cost delivery systems, such delivery systems cannot work ideally for green construction projects because green materials, products, and technologies are often unique and expensive. In addition, the interviewees expressed the view that sound communications between design and construction teams are required to make for a successful project, while the least-cost delivery systems, such as design-bid-build, are unable to achieve that level of communication because two separate teams manage the design and construction.

5.3.2 Obstacles Encountered in Green Construction Projects

The survey results revealed that all the respondents encountered some obstacles during the management of green construction projects. Table 5.1 presents the type of obstacles listed in the survey and the frequency of respondents who have encountered them in practice.

According to the survey results and as underlined in Table 5.1, the top five obstacles encountered by professionals and managers when managing a green construction project are: (1) the high premium cost associated with green construction; (2) the lack of communication and interest between project members; (3) the lack of expressed interest from client or market demand; (4) the lack of credible research on the benefits of green buildings; and (5) green building practices are costly to implement. Surprisingly, none of the respondents felt that there is a lack of expertise and knowledge in green building and its principles. This could indicate that, despite a lack of green product information for the developer, there are ample consultancy services provided in Singapore to assist in green construction. Furthermore, none of them felt that there was a lack of government support for green construction, which could be due to the tremendous effort that the government-based BCA has expended to promote the development of green building and sustainable construction. Moreover, through the Green Mark Incentive Scheme (GMIS), the adoption of green building design, construction practices, and environmentally friendly technologies, have been hastened.

5.3.3 Impacts of the Obstacles on Green Construction Projects

The respondents were also requested to state the extent of the effects that these obstacles have on the main objectives in green construction projects, which are project schedule, budget, and quality of the end product. They were asked to rate the magnitude of the effects on a scale of 1–5, where 1 represents no effect on the objectives at all and 5 represents an extremely large effect. As summarized in Table 5.2, the impact the obstacles have on keeping up with project schedule and quality of the end product have mean scores of 3.32 and 3.16, respectively, and that the effect of obstacles on the project budget has a higher mean score of 4.10, suggesting that the obstacles have a more extensive effect on project cost. Moreover, the standard deviation of its mean score is relatively lower than the other two objectives, which indicates that opinions were not as diverse. This suggests that most GMPs and GMMs share a common impression that obstacles encountered during management of green construction projects can affect project cost extensively. To further test the accuracy of the results, the objectives, namely, schedule, budget and quality, were paired with each other to carry out a paired samples *t*-test, as shown in Table 5.2. In pair 1, the effect of obstacles on the project schedule is compared to the effect on the project budget. In pair 2, the effect of obstacles on the project schedule is compared to the effect on building quality, and in pair 3, the effect of obstacles on project

TABLE 5.2 Summary of Impacts of Obstacles on Green Project Objectives

Objectives	N	Mean	Standard Deviation	Pairs for *T*-Test	*P*-Value	Mean Difference Significant? ($P<.10$)
Schedule	31	3.320	0.871	Schedule vs Budget	.000	Yes
Budget	31	4.100	0.539	Schedule vs Quality	.057	Yes
Quality	31	3.160	0.820	Budget vs Quality	.000	Yes

budget is compared to the effect on building quality. This test is carried out to see if there are any statistically significant differences between two samples at a confidence level of 0.10. The results showed that there are significant differences between the means of all of the pairs and that this can be interpreted to mean that the obstacles mentioned in the previous section indeed impact the project cost the most.

Similarly, according to the interviewees, the most significant impact of the obstacles is on the project budget. Expensive technologies, products, materials, as well as consultancy services that come with green buildings, take a toll on the construction budget. To minimize the adverse impact of these obstacles, local project managers have taken various measures, and one of the common measures they take is to develop cost estimates as close to the actual sum as possible during the project planning stage. Additionally, they are also searching for some alternatives that are economical. However, searching for alternatives requires a tremendous amount of time, which is often a luxury in almost all construction schedules. A balance has to be achieved to ensure that cheaper alternatives are sourced while keeping up with the schedule. Like any other construction project, delays in the construction schedule can result in copious losses. Furthermore, the obstacles can also exert an adverse impact on the quality of the project. This is because, as a cost saving measure, the project team would sometimes use substandard products and materials, which are not necessarily sustainable or environmentally friendly, thus compromising the energy efficiency of the project.

5.3.4 Solutions for the Obstacles

Through the survey, the respondents were also asked to choose solutions that can help overcome the obstacles listed in Table 5.1. The percentage of respondents who chose particular solutions for each of the obstacles is tabulated in Table 5.3. Obstacles 3, 4 and 9 are excluded in this section because none of the respondents consider them as obstacles encountered in green building project management. Only the top five obstacles (1, 2, 6, 10, and 12) mentioned in the previous section are analyzed in detail here.

Both of the obstacles involving high cost premium and costly green construction practices are cost related and are considered to be the biggest obstacles a project management team has to overcome. This factor has an extensive effect on the projects' budgets because, in Singapore's profit driven construction industry, most projects are awarded based on the lowest tender price. Hence, a green construction project will be unattractive in comparison to conventional construction projects regarding cost. Moreover, BCA's GMIS is capped at an amount of $20 million. According to one of the survey respondents, the fund has only $7 million left at this point. Therefore, with a tighter budget, future projects may only aim to attain the minimum green requirements necessary to comply with local regulation. To overcome the problem of the high cost

TABLE 5.3 Summary of Suggested Solutions for Obstacles

Obstacles	Solutions	Frequency	%
1 and 10	Interest free lending schemes provided by government to overcome market and financial barriers	8	25.8
	Educating owners on the future benefits of green buildings	20	64.5
	Insistence from client	7	22.6
	Green Mark Certification to be made mandatory for all new and existing buildings by authority	3	9.7
	Government to provide incentives to offset high premiums of green building projects	21	67.7
	Advocated by professionals such as architects and engineers	0	0.0
	Public and market demand for green buildings	15	48.4
2	Conduct tool-box meeting for regularly	23	74.2
	Engaging personnel with green building background	14	45.2
5	A green building practice framework which is simple enough for project managers and other professionals to follow	9	29.0
	Decrease the amount of premium provided for the project if green construction practices are not adopted	5	16.1
	Insistence from client	8	25.8
	Heavier taxes and penalties on unsustainable construction practices	5	16.1
6	Research studies or evidence to show that green building help increase productivity and health of occupants	4	12.9
	Organize construction tour to introduce and educate the public about the benefits of green building	25	80.6
7	Bonuses provided for staff if the building is green mark certified or qualified for green mark awards	27	87.1

Continued

TABLE 5.3 Summary of Suggested Solutions for Obstacles—cont'd

Obstacles	Solutions	Frequency	%
8	Bonuses provided for staff if the building is green mark certified or qualified for green mark awards	7	22.6
	Project team to create a culture for the adoption of green building practices	6	19.4
	Create a simple green building practice framework which employees can follow easily to facilitate the transition from conventional to green practices	13	41.9
9	Subsidy or incentives from government for green building projects	20	64.5
	Guaranteed enhancement to company's reputation through publication and certifications	5	16.1
11	Engaging personnel with green building background	28	90.3
	Training provided on green building before the start of project	7	22.6
	Upgrading courses to supplement knowledge regarding sustainable construction	7	22.6
12	Subsidy from government for R & D in green building systems and management	30	96.8
13	Summarize these codes and regulations into a simple checklist, which can be comprehended and followed easily	22	71.0
	Introductory courses on green building legislations conducted for all staff members before the start of project	4	12.9

involved in green construction, all respondents felt that government incentives for green construction projects can help offset the high cost involved. 67.7% of the respondents felt that educating the client on the future benefits of green construction could be the solution to this problem as well. With clients internalizing the potential benefits reaped from green construction, a high cost premium may represent less of a deterrent to the client.

Because green construction is a fairly new concept in Singapore's construction industry, it is important to communicate the green goals and objectives to all stakeholders and project team members to achieve successful project execution. To solve the problem of lack of communication between project members as an obstacle in green building project management, 74.2% of the respondents felt that a regular toolbox meeting could be conducted to ensure that important information about the project is well communicated. Engaging personnel with green building experience was cited by 45.2% of them as a means to overcome this obstacle as well.

The lack of client interest in green building and market demand from the public is another significant obstacle in green building project management. The client plays a major role in creating and leading an efficient and focused project team. Without expressed interest from an enlightened client, it is difficult for the project team to execute green construction practices and green design features. From the survey, 80.6% of the respondents felt that a construction tour could be organized for the client and the public to educate them on the benefits of green building so as to increase their interest and create a higher market demand for green buildings in Singapore.

The poor demand for green buildings could also be due to the lack of credible research on their benefits. Knowing the advantages that green buildings could bring about heightened interest and 96.8% of the respondents felt that a subsidy from the government for research and development of green building systems and management could provide concrete evidence of how beneficial they are to humans, society, and the economy.

Results of the postsurvey interview show that several measures have been taken by the local industry to cope with the top five obstacles identified by the survey questionnaire. To keep a green construction project within budget despite the high cost involved, normally, the project team develops the building in terms of the most basic Green Mark requirement only, unless the client wishes to seek a higher-level certification, such as Green Mark Gold or Platinum. Also, cheaper material alternatives are sourced to keep the material cost low. Prices from different subcontractors are compared and the one with the lowest price will be chosen, provided that the products or systems meet certain quality standards. These standards are determined by how energy efficient the green building has to be, as the materials, products, and systems used in the building determine how much energy will be consumed. Having in-house consultants also helps to eliminate the cost of engaging third party consultancy services, especially for bigger projects. In fact, many personnel involved in green building construction play multiple roles. For instance, some engineers are trained under the Energy Sustainability Unit's (ESU) Locally-based Enterprise Advancement Program (LEAP) as energy managers, who are capable of carrying out an energy audit for a building. Thus, the cost is reduced as the need to engage external energy auditing services is avoided. Also, professionals and managers with experience in green building and familiar with the local sustainability legislation are preferred and selected to be in the project team. By having multidisciplinary personnel, who are familiar with green building and the relevant legislation, involved in the project team, communication between individuals can be enhanced at the same time.

To overcome the top five obstacles stated earlier, additional possible solutions are proposed in this section. These solutions are developed based on a literature review of the drivers behind the growing trend in green building, results from the survey questionnaire, and interviews.

5.3.4.1 Providing Governmental Incentives for Green Construction Projects

To overcome the cost issue associated with green buildings, it is recommended that the government continues to incentivize the adoption of green construction until an adequate market demand for green buildings is achieved. The scope of incentives should be expanded from developers, building owners, architects, and mechanical and electrical consultants, so as to cover the use of green products and technologies, such as solar panels, which are still very expensive now. Doing so helps to keep the cost of building green within a reasonable and acceptable budget, which could effectively remove deterrence and facilitate the adoption of green designs and practices.

Demand for green buildings can be generated by educating the public about the importance of sustainable development due to depleting world resources and deteriorating global climate. This can be done by organizing green building construction tours for building owners and potential users. In addition, government agencies can educate the public about the benefits of green building and construction practices. Similar to the one given by a builder in Michigan, the construction walkthrough should be opened to the public before air-conditioning and mechanical ventilation systems are installed (Builder, 2009), so that the visitors can appreciate how green building designs and technologies can help make buildings energy efficient and sustainable. The tour should update the public and developers about the breakthroughs in building approaches, mechanical and electrical systems, such as efficient energy ventilation and low-energy lighting systems, energy conservation procedures, and the overall process of green building, including sustainable construction methods and site planning (Builder, 2009).

5.3.4.2 Improving Communication Within the Project Team

For a green construction project to be successful, a higher level of communication is required among project team members compared to conventional building projects. This is because it is crucial for every team member to have an in-depth understanding of green building principles and of their obligation to contribute with ingenuity and vigor to produce an outstanding building. Thus, other than conventional solutions, such as regular toolbox meetings, the clients should advocate green management goals from the top down to motivate the whole project team toward the goal. Also, the project team's communication can be enhanced by engaging personnel with green building expertise as part of the team. BCA's certified GMM and GMP scheme is a useful way of acknowledging managers and professionals who are experienced in green building.

The project managers of green construction projects may also implement high-performance green building delivery systems instead of conventional delivery systems, which may not be suitable for green construction projects. According to Kibert (2008), this delivery system is derived from the construction management at risk delivery method, and intense communication between project members is essential. Correspondingly, inaugural team building should intermesh a wide variety of stakeholders to ensure that everyone understands the project's targets and green specifications (Kibert, 2008). The delivery system also requires all personnel to be proficient in green building (Kibert, 2008). Because green buildings in Singapore have to attain Green Mark Certification, project members have to be very familiar with the Building Control (Environmental Sustainability) Legislation 2008 and Green Mark requirements. This system seeks to minimize adversarial relationships and simplify negotiations among the parties (Kibert, 2008).

To ensure that the project management team comprises people who have good knowledge about green buildings, the client or project manager could request the qualifications of team members during the selection process. As mentioned by Kibert (2008), the selection can be based on experience, qualifications, previous work, and understanding of the owner's program and requirements, the building site, and the firm members' ability to work with other project team members.

5.3.4.3 Conducting Research on Benefits of Green Building

As expressed by survey respondents, one of the reasons behind the poor market demand for green building in Singapore is the lack of credible research on the benefits of green buildings. Therefore, it would be useful for government agencies to provide a subsidy for the research and development of green building systems in Singapore. Moreover, some green technologies, such as photovoltaic, are still not commercially viable and it would be helpful for the government to provide funds to bridge the research and development gap in order to bring these technologies to fruition.

5.3.5 Green Construction Project Management Framework

During the postsurvey interviews, all the interviewees stated that they do not have a specific project management framework for green building construction projects. Therefore, a conventional project management framework was being used as a guideline for sustainable construction in Singapore. However, special attention is being paid to laws and regulations that involve green building requirements. Requirements stated in BCA's Green Mark Awards are also analyzed before the commencement of construction to determine if attainment of a certain level of the award is feasible and within the project team's capabilities, availability of resources, and most importantly, cost.

5.4 SUMMARY

This chapter investigated the common obstacles encountered during the management of green construction projects and proposed feasible solutions to overcome these obstacles. This chapter showed that the main dissimilarity between conventional and green construction projects is the level of details and communication required. This is especially reflected in the delivery systems adopted. Conventional construction projects typically adopt least-cost delivery, such as design-bid-build, in which communication between the design team and the construction team is not prioritized. In addition, a succinct schematic design is usually adequate at the planning stage of conventional construction projects. Comparatively, green construction projects require superlative communication, which can only be achieved through delivery systems like design-build, in which building design and construction is carried out as a single entity. Furthermore, the detailed integrated design process is employed at the start of the project, because, unlike conventional building design, green design features are unique and require deep integration with every aspect of the building.

Results from the survey and interviews revealed profound obstacles in the project management of green construction projects. These obstacles were found to be interrelated, but they ultimately boil down to the high cost premium of green building. The lack of research and development on the benefits of green buildings and green technologies is a driver behind the lack of demand for buildings to go beyond the legal requirement. As such, these green technologies and systems are not prevalent, leading to the hefty price tags associated with their installation and implementation. A vicious cycle starts, making green construction practices costly to implement. This deters clients and building owners from building green and adopting sustainable construction practices beyond their legal obligation. Also, green construction requires collaborative effort and communication between the project team members to succeed. However, many local projects adopt a conventional design-bid-build form for management, in which the design and construction are managed as separate entities, causing a lack of communication between team members. These factors lead to an onerous project that is more difficult to manage than a conventional one, thus discouraging owners to take up a green building project.

To overcome these obstacles and bring success to green construction projects, there are several imperative solutions can be considered. They include expanding fiscal measures whereby the government provides incentives for professionals as well as research and development and use of green products and technologies; providing the public with education on green building advantages to drive demand; adopting communication-conducive delivery systems, such as the high performance green building delivery system that allows design and construction to be integrated. Through the proposed solutions, the construction industry may be keener to undertake green construction instead of conventional construction in the future. In addition, the demand for green buildings is expected to increase along with public awareness of environmental issues and the advantages of sustainable construction. Such public awareness will increase as a result of improved research and development in green products and technologies, which will yield credible research on the benefits of green buildings.

Chapter 6

Knowledge Areas and Skills for Green Construction Project Management

6.1 INTRODUCTION

A competent project manager is vital to project success. According to Frank (2002), the project manager has a direct influence over 34%–47% of project success. Ahadzie (2007) also confirmed the industry's growing awareness of the relationship between achieving project success and construction project management competencies. Therefore, successful construction organizations now focus on ensuring that project managers acquire the core competencies required to be successful in their assignments.

An organization can maximize its probability of consistently attaining project success by recruiting, developing, nurturing, and retaining superior project managers. Working closely with a range of other professionals, project managers organize, plan, schedule, and control the work, and are responsible for getting the project completed within the established time and cost limitations. To manage the project professionally and successfully, a project manager needs to possess the required skills and knowledge. Widespread research studies have documented the managerial skills necessary for efficient project performance (Smallwood, 2006). Given the fast-changing environment of the construction industry, with challenges such as skills shortages, the rapid advancement of information and communication technologies, and the increasing prioritization of issues such as sustainability, environmental protection and climate change, the role of the project manager needs to be adapted.

With mounting global concern for the negative impacts on the environment caused by human activities in recent years, many industries are turning toward sustainable development and implementing green measures. Building construction companies from various regions around the world have integrated green concepts into their construction plans to mitigate the impacts on the environment (Hwang and Tan, 2012). As the industry changes, project managers find themselves confronted by new issues and must undertake roles that have not traditionally been part of their responsibility (Edum-Fotwe and McCaffer, 2000). Both Ceran and Dorman (1995) and Russell et al. (1997) recognized

Performance and Improvement of Green Construction Projects. https://doi.org/10.1016/B978-0-12-815483-0.00006-5

the changing role of construction project managers and argued that they must supplement their traditional functions with nonengineering knowledge and skills to meet today's professional demands. Today's project manager fulfills not only traditional roles of project management but also must manage the project in the most efficient and effective manner with respect to sustainability.

As the green building construction phenomenon continues to grow and gain popularity, there is a need to better understand the pivotal attributes that project managers should possess to manage green construction projects. Despite numerous studies on a project manager's competency, few have specifically examined what critical knowledge and skills are required for the project manager to successfully deliver a sustainable project. As a result, with the intent to enhance sustainability efforts without compromising the competitiveness of project managers, this chapter aims (1) to identify the essential knowledge and skills required to be a competent project manager of green construction projects; (2) to discover the challenges that project managers encounter in managing green construction projects and determine critical knowledge areas and skills that can respond to the challenges; and finally, (3) to provide a comparison of critical knowledge areas and skills between conventional and green construction projects.

6.2 METHOD AND DATA PRESENTATION

Having conducted a comprehensive literature review on the essential knowledge and skills required for competent project managers who execute green projects, a presurvey questionnaire was developed. The presurvey questionnaire aimed to validate 39 knowledge and skill areas that had been identified through the literature review, as listed in Tables 6.1 and 6.2. Using the mean value ranking method, the top 20 knowledge and skill areas were selected and eventually adopted for the main survey. Ultimately, a total of 52 completed survey questionnaires were received, tabulated and analyzed. In summary, the distribution of the respondents' years of experience in project management was found to be as follows: up to 5 years (11.5%), 6–10 years (26.9%), 11–15 years (21.2%), 16–20 years (26.9%) and above 20 years (13.5%).

In addition, interviews with industry experts were conducted to fine-tune the list of challenges that project managers face in managing green construction projects. The interviewees discussed the challenges identified in the literature review, as listed in Table 6.3. These results enhanced the understanding of the rationale for the challenges that were to be included in the questionnaire survey. Three project managers from different companies were engaged for interview, all of them having more than 10 years of experience in managing green construction projects.

The purpose of the main survey was (1) to discover the challenges that project managers encounter in green construction and (2) to identify critical attributes that project managers need to possess in managing green construction

TABLE 6.1 Summary of Knowledge Areas

No.	Essential Knowledge Area	Project Management Institute (2008)	Dogbegah et al. (2011)	Odusami (2002)	Gushgari et al. (1997)	Kerzner (2013)	Ling (2003)
1	Schedule management and planning	√	√	√	√	√	√
2	Cost management	√	√				√
3	Quality management	√	√	√	√		√
4	Human resources management	√	√	√		√	
5	Risk management	√	√				
6	Supply chain management	√	√				
7	Claims management		√				
8	Knowledgeme management		√				√
9	Health and safety management		√				√
10	Conflict and dispute management		√			√	
11	Ethical management		√				

Continued

TABLE 6.1 Summary of Knowledge Areas—cont'd

No.	Essential Knowledge Area	Project Management Institute (2008)	Dogbegah et al. (2011)	Odusami (2002)	Gushgari et al. (1997)	Kerzner (2013)	Ling (2003)
12	Stakeholder management		√				√
13	Information technology management		√				√
14	Communication management		√	√	√		
15	Materials resources management		√			√	
16	Financial management		√	√	√		
17	Plant and equipment resources management		√			√	

TABLE 6.2 Summary of Skills

No.	Essential Skills	Edum-Fotwe and McCaffer (2000)	Odusami (2002)	Gushgari et al. (1997)	Fraser (1999)	Tett et al. (2000)
1	Basic technical skill	√	√	√	√	
2	Site layout and mobilization	√			√	
3	Estimating and tendering	√			√	
4	Design activities and background	√			√	
5	Reading and understanding drawings	√			√	
6	Technical writing	√			√	
7	Leadership	√	√	√		
8	Decision making	√	√	√		√
9	Problem solving	√	√	√		√
10	Negotiation	√	√	√		
11	Human behavior	√				
12	Delegation	√	√	√	√	√
13	Team working	√	√		√	
14	Stress handling	√	√			

Continued

TABLE 6.2 Summary of Skills—cont'd

No.	Essential Skills	Edum-Fotwe and McCaffer (2000)	Odusami (2002)	Gushgari et al. (1997)	Fraser (1999)	Tett et al. (2000)
15	IT skills	√				
16	Drafting contracts	√			√	
17	Presentation	√				√
18	Report writing	√				√
19	Public speaking	√				√
20	Marketing and sales	√				
21	Chairing meetings	√			√	
22	Public relation	√			√	√

TABLE 6.3 Challenges faced in green construction projects

No.	Challenge	Zhang et al. (2011)	Tagaza and Wilson (2004)	Hwang and Tan (2012)	Eisenberg et al. (2002)
1	Higher costs for green construction practices and green material	√	√	√	
2	Technical difficulty during the construction process	√	√		
3	Risk due to different contract forms of project delivery		√	√	
4	Lengthy planning and approval process for new green technologies and recycled materials	√	√		√
5	Unfamiliarity with green technologies	√	√		√
6	Greater communication and interest is required among project team members		√	√	
7	More time is required to implement green construction practices onsite		√	√	

projects. The results from the presurvey and interviews served as a platform in developing the main survey questionnaire. The first section of the survey questionnaire captured the respondent's profile. The second section listed the challenges that project managers face in managing green construction projects. The respondents were asked to rate the extent to which each of the challenges affected them using a 5-point Likert scale (1 = not relevant at all; 5 = most relevant). The next section asked the respondents to rate the importance of project manager attributes in responding to the different challenges identified in the second section of the survey (1 = not important at all; 5 = most important). The last section requested the respondents to rate the relative importance of the attributes in comparison to conventional construction projects. This section aimed to identify the importance of each project management knowledge area and skill that a project manager has to strengthen in order to be competent in a green construction project when compared to conventional projects.

Out of the 500 questionnaires sent out, 30 sets of completed survey questionnaires were received and analyzed. The response rate was relatively low because only the responses from the respondents who had experience in managing green construction projects were considered valid. All the respondents had more than 4 years of experience in managing conventional projects and had some experience with green construction projects (up to 1 year, 7%; 1–2 years, 17%; 2–3 years, 23%; 3–4 years, 23%; more than 4 years, 30%). The data also captured the type and nature of a total of 68 green construction projects managed by the respondents throughout their careers. There were 23 commercial projects (34%), 19 residential projects (28%), and 26 educational projects (38%). The majority (84%) of the projects recorded from the survey involved new construction, with the remaining (16%) being addition and alteration projects.

6.3 ANALYSIS RESULTS AND DISCUSSIONS

6.3.1 Critical Challenges in Green Construction Projects

The following subsections discuss the analysis results on various challenges that project managers face during green construction project management. Table 6.4 summarizes the results with specific challenges categorized into seven areas, their mean scores, and ranks within and across the categories.

6.3.1.1 Planning-Related Challenges

Most challenges faced by project managers executing green construction projects are planning related. The longer time required for the preconstruction process, P6, yielded the highest mean score of 4.57, indicating that it is the most frequently encountered challenge. The difficulty in comprehending green specifications given in the contract details, P7, was ranked second highest, while the lengthy approval process to approve new technologies, P5, was ranked as the least critical planning-related challenge, with a mean value of 2.43.

TABLE 6.4 Analysis Summary: Challenges

Category	Challenge	Code	Mean	Rank Within Category	Rank Cross Category
Planning-related	Adoption of different contract forms of project delivery	P1	3.20	6	28
	The design, orientation and structure of the building	P2	3.33	5	27
	Planning of different construction sequence	P3	4.20	3	8
	Planning of different construction technique	P4	4.13	4	10
	Lengthy approval process for new green technologies within the organization	P5	2.43	7	36
	Longer time required during the preconstruction process	P6	4.57	1	1
	Difficulty in comprehending the green specifications in the contract details	P7	4.27	2	7
Project-related	Difficulty in approving payment disbursement to suppliers and subcontractors	T1	3.07	4	33
	Difficulty in assessing the progress of completion in green construction	T2	2.97	5	34
	Difficulty in the selection of subcontractors in providing green construction service	T3	4.50	1	2
	More time is required to implement green construction practices onsite	T4	3.97	3	15
	More alteration and variation with the design during the construction process	T5	4.37	2	6

Continued

TABLE 6.4 Analysis Summary: Challenges—cont'd

Category	Challenge	Code	Mean	Rank Within Category	Rank Cross Category
Client-related	Specific budget specification of the green project	C1	3.77	4	20
	Objective of the building project	C2	4.01	1	14
	Required date of completion	C3	3.67	5	21
	Level of risk the client is willing to take in green technologies	C4	3.97	2	16
	Client uses a lot of time in making decision	C5	3.63	6	22
	Special request from client pertaining to specified green technologies to be used	C6	3.93	3	18
Project team-related	Conflict with the architect over the type of material to be used	PT1	3.17	4	29
	Lack of communication and interest among project team members	PT2	4.10	2	11
	Frequent meetings with green specialists	PT3	4.37	1	5
	Green consultant delay in providing information	PT4	3.13	5	30
	Conflict of interest between consultant and project manager	PT5	3.53	3	24
	Specific performance required for green building projects	PT6	2.53	6	35

Material and equipment related	High cost in green material and equipment	M1	4.43	2	4
	Uncertainty with green material and equipment	M2	4.50	1	3
	Availability of green material and equipment	M3	4.01	4	13
	Decision on different green material and equipment	M4	4.07	3	12
	Imported green material or equipment	M5	3.97	5	17
Labor-related	Resistance to change from their traditional practices	L1	3.60	1	23
	Lack of the technical skill regarding green technologies and techniques	L2	3.37	3	26
	Workers' unaware of the correct methods and procedures	L3	3.50	2	25
External	Government policy	E1	3.87	2	19
	Lengthy BCA approval process for new technologies	E2	3.10	3	31
	More time required for TOP due to green mark certification process by BCA	E3	3.07	4	32
	Unforeseen circumstances in green project	E4	4.20	1	8

According to Kubba (2010), over time green construction projects progressively incorporate more advanced and intricate systems. During design, the impact of the elements on each system must be considered as a whole. A failure to take into account the integration of green technologies and their impact on other building elements results in construction conflicts, leading to delay in order to address such problems. The interviewees verified that the preconstruction period of a green construction project usually requires more time as compared to conventional construction projects. The reason for this is the need for a more detailed and comprehensive design plan, which incorporates all the green features required by the construction project. This often means more involvement and interaction with different stakeholder groups, which may also slow down the preconstruction process. The interviewees emphasized the importance that during the design development stage, the design team needs more time to study the design system performance against the stated green building goals. Most green technologies are relatively new to the construction industry and their system performance is not well known. Furthermore, greater efforts are needed to ensure that the green features that are incorporated into the design incorporate cheaper, but still efficient green options. Furthermore, the design, orientation, and structure of the building (P2) must play a key part in selecting the most effective green features to be employed in a given building. For instance, solar panels should be located in places where there is maximum sunlight. Although green construction projects often have green consultants and green engineers, the project manager must have the ability to assess the factors at hand and arrive at the best solutions for the project.

6.3.1.2 Project-Related Challenges

Among all the listed challenges in the category of project-related challenges, T3, difficulties associated with the selection of subcontractors who provide green construction, had the highest mean of 4.50.

Selection of subcontractors for traditional construction projects may be less difficult for experienced project managers who are familiar with the performance of subcontractors with whom they have worked in the past. It is also likely that there is insufficient performance information for subcontractors who are involved in green construction projects. As a result, the selection process becomes tougher and may take more time. As for the second ranked challenge, which identified alterations and variations associated with the design during the construction process as a significant issue, one of the interviewees suggested that this element ranks highly because there seemed to be more such changes in green construction projects. One possible explanation is the inconsistency of the drawings provided by the architect and the ultimate design that incorporates the green features. To respond to this challenge, project managers must coordinate more intensively with stakeholders in order to deliver the project successfully.

6.3.1.3 Client-Related Challenges

As shown in Table 6.4, understanding the objective that the client wants to achieve with a green building, C2, had the highest mean of 4.01. One interviewee emphasized the importance for project managers to understand the objectives of their clients, especially when the client wants to achieve a certified green mark. In order to achieve the goal, the client should specify the type of green technology to be used, although this too may pose a challenge if the specified green technology is not available locally. Also, the level of risk that the client is willing to undertake can be another challenge for the project manager. Project managers are responsible for ensuring that the green technologies applied in the project are safe, especially if the client has a low threshold for risk. In some cases, green consultants may try to convince the client to adopt a particular system. The client may need more than the usual amount of time to make decisions on whether or not to implement such a system.

6.3.1.4 Project Team-Related Challenges

The analysis results in Table 6.4 show that PT3, frequent meetings with green specialists, is one of the most critical project team-related challenges faced by project managers. An interviewee confirmed that there are an increased number of meetings required for a green construction project because an intense alliance with green specialists is necessary to fine-tune emerging issues. Some interviewees also noted that necessary information from specialists tended to be delayed (PT4), affecting the entire project schedule. Furthermore, despite the need for closer collaboration within project teams, project managers reported issues like lack of communication and interest among project team members (PT2).

6.3.1.5 Material and Equipment-Related Challenges

In terms of material and equipment-related challenges, the respondents cited uncertainties with green material and equipment, M2 (Mean = 4.50), as the most relevant challenge that they faced. All interviewees expressed concern over the reliability of green materials and equipment as most of these new technologies do not have a proven track record. The second highest challenge cited was the cost of green materials, M1. Research done by Zhang et al. (2011) indicated that using green materials would cost from 3% to 4% more than conventional construction materials. Selection of the green materials has to be cost-conscious in order to prevent budget overruns. Unlike conventional construction materials, many green materials may not be available locally. Problems arising from imported material may include understanding the legal regulations of the various countries. Furthermore, to ensure that an imported material is compatible for local use, extensive testing may be required.

6.3.1.6 *Labor-Related Challenges*

In the category of labor-related challenges, workers' resistance to changing their traditional practices (L1) was the most critical challenge faced by project managers. Although the project managers may not directly deal with the workers on site, their work plays a significant role in the success of the project. If the workers lack the necessary technical skills (L2) or are unaware of the correct procedures (L3), they may have a negative impact on achieving project success.

6.3.1.7 *External Challenges*

External challenges can impact a project in many ways. Unforeseen circumstances (E4) may be even less predictable in green construction projects as compared to conventional construction projects. Green construction projects are still relatively new in Singapore, and consequently, the team members and workers have little experience. When faced with unforeseen circumstances, project managers may not be equipped with any prior knowledge to handle the situation. One interviewee quoted an instance in which the client had decided to aim for a higher green mark score, switching from gold to platinum, in the middle of the construction project. This unforeseen change created a considerable challenge for the project managers in rescheduling the construction process according to the new design and requirements. Understanding governmental policies related to green construction projects (E1) is also vital and was ranked highly. Strict policies and regulations are designed to protect human health and the environment, and a failure to comply may cause project delay, termination, and fines. Even if green consultants are involved during most of the green construction project, it is paramount that project managers have a good understanding of the policies.

6.3.2 Knowledge Areas and Skills to Overcome the Challenges

This section presents the results of the survey analysis. Knowledge and skill areas that project managers need in order to effectively deal with the challenges of green construction are discussed. Tables 6.5 and 6.6 list the knowledge and skills that were identified through literature review and verified by the presurvey process.

6.3.2.1 *Knowledge Areas and Skills for the Planning-Related Challenges*

As shown in Table 6.5, the survey responses revealed that the top three knowledge areas critical to project planning-related challenges are (1) schedule management and planning; (2) communication management; and (3) risk management. As discussed, more time is usually required during the preproject planning process for green projects. Therefore, project managers must be effective at

TABLE 6.5 Analysis Summary: Critical Management Knowledge Areas

Management Knowledge Area	Category of Challenges													
	Planning-Related		*Project-Related*		*Client-Related*		*Project Team-Related*		*Material and Equipment-Related*		*Labor-Related*		*External*	
	Mean	*Rank*	*Mean*	*Rank*	*Mean*	*Rank*	*Mean*	*Rank*	*Mean*	*Rank*	*Mean*	*Rank*	*Mean*	*Rank*
Schedule management and planning	4.37	1	4.53	1	4.17	5	3.17	9	3.00	9	3.53	3	4.00	1
Health and safety management	3.93	5	4.07	4	3.60	7	3.23	8	3.50	6	2.80	7	2.70	7
Human resources management	3.60	8	3.60	9	3.53	9	3.53	5	3.37	8	4.40	1	3.07	5
Stakeholder management	3.80	7	3.87	7	4.37	1	4.03	3	3.83	4	3.23	6	3.23	4
Cost management	3.83	6	4.30	3	4.23	2	3.43	6	4.67	1	2.67	8	2.60	8
Communication management	4.27	2	3.87	6	4.20	3	4.30	1	3.53	5	4.00	2	3.63	2
Conflict and dispute management	4.03	4	3.93	5	3.73	6	4.17	2	3.43	7	3.53	4	2.83	6
Risk management	4.17	3	4.33	2	4.20	4	4.03	4	4.03	3	3.50	5	3.50	3
Materials resources management	3.57	9	3.70	8	3.57	8	3.23	7	4.40	2	2.43	10	2.47	10
Claims management	3.13	10	3.23	10	3.27	10	2.90	10	2.87	10	2.67	9	2.60	9

TABLE 6.6 Analysis Summary: Critical Management Skills

	Category of Challenges													
	Planning-Related		*Project-Related*		*Client-Related*		*Project Team-Related*		*Material and Equipment-Related*		*Labor-Related*		*External*	
Management Skill	*Mean*	*Rank*	*Mean*	*Rank*	*Mean*	*Rank*	*Mean*	*Rank*	*Mean*	*Rank*	*Mean*	*Rank*	*Mean*	*Rank*
Delegation	4.37	2	4.47	2	3.93	5	3.60	8	3.73	5	4.40	1	2.83	8
Leadership	3.67	8	3.77	7	4.23	2	3.93	3	3.40	6	4.07	2	3.07	5
Decision making	4.33	5	4.53	1	3.83	6	3.83	6	4.30	1	3.07	7	3.30	3
Problem solving	4.37	4	4.33	3	4.17	3	3.73	7	3.83	3	3.17	6	4.10	1
Team working	4.20	6	4.10	6	3.77	8	4.27	1	2.87	7	3.47	3	2.87	7
Analytical	4.50	1	4.33	4	4.50	1	3.53	9	4.07	2	3.00	8	3.50	2
Presentation	3.50	9	3.07	10	3.17	10	3.23	10	2.30	9	2.23	10	2.10	10
Human behavior	3.33	10	3.57	8	3.80	7	3.87	4	2.50	8	3.33	4	3.03	6
Negotiation	4.37	3	4.20	5	4.17	4	4.10	2	3.77	4	3.23	5	3.20	4
Chairing meetings	3.70	7	3.47	9	3.27	9	3.83	5	2.30	10	2.37	9	2.27	9

schedule and planning management to ensure that the project is completed on time. An interviewee suggested that a balance has to be achieved to ensure that cheaper alternatives are sourced while keeping up with the schedule. Furthermore, sequencing for green construction may require the project managers to be more detailed when planning the project. Communication management is also important as project managers have to ensure that information is communicated effectively to all the various parties involved in the preproject process including green specialists and architects.

Analytical skill was found to be the most important skill required to address preproject-related challenges, as shown in Table 6.6. An interviewee mentioned that it is crucial for the project managers to analyze the whole situation and come up with an integrated project schedule that is achievable and manageable.

6.3.2.2 Knowledge Areas and Skills for the Project-Related Challenges

Table 6.5 shows that for project-related challenges, the respondents ranked schedule management and planning as the most important knowledge area, followed by risk management. Risk management is particularly important because it can assist in the selection of the subcontractors who provide technologies and techniques required for green construction. Project managers have to be aware of the risks involved and careful consideration must be given to the selection of a new subcontractor. Nonetheless, a longer risk assessment and decision-making process can cause delay. Additions and alterations to the construction scope may also affect the planned project schedule. Project managers need to be equipped with good scheduling and planning knowledge to cope with these challenges. The most important skill to manage project-related challenges was found to be decision-making with a mean score of 4.6 (Table 6.6). This result could be explained by the need for project managers to make the best possible decision on the selection of specific technologies, systems, and subcontractors required for green projects.

6.3.2.3 Knowledge Areas and Skills for Client-Related Challenges

Table 6.5 shows that stakeholder management is ranked as the most important knowledge area to address the client-related challenges. As confirmed by the interviewees, this result may be explained by the importance of understanding the client's objectives. Clients are considered to among the stakeholders in a project and the ability to manage them, by understanding their needs and objectives, is essential to project success. Cost management was also ranked highly. One interviewee mentioned that it is his company's practice to keep a green building construction project within budget despite the high cost involved, but this was accomplished by achieving the most basic Green Mark requirement and fulfilling the minimum Building Control (Environment Sustainability) Regulations, unless the client requested a higher certification level, such as

Green Mark Gold or Platinum. It could be argued that stakeholder management and cost management can be equally important as they are applied simultaneously to react to client-related challenges.

Analytical skill was reported as the most important skill for managing the client-related challenges. This could imply that competent analysis is crucial to analyze and manage stakeholder requirements and to come up with the best solutions.

6.3.2.4 Knowledge Areas and Skills for Project Team-Related Challenges

Knowledge of communication management was ranked first with a mean of 4.30. Because green projects require a more holistic and integrated approach, the design process is more complex and the design phase often overlaps with construction (Glavinich, 2008). Some of the interviewees reported the need for everyone in the project team to be well informed of the project's goals and objectives in order to prevent errors. Conflict and dispute management was ranked second, as shown in Table 6.5. Conflict with other project team members may be unavoidable and project managers must be equipped to effectively manage conflict without affecting the progress of the project.

The survey results identified teamwork as the most important skill in this category (Table 6.6). Because challenges in this category are related to the project team, it is reasonable that project managers who are equipped with good team building skills could enhance team cohesiveness, improving the overall team performance.

6.3.2.5 Knowledge Areas and Skills for Material and Equipment-Related Challenges

Based on Table 6.5, knowledge of cost management (mean = 4.67) was ranked first for dealing with material and equipment-related challenges, followed by material resources management (mean = 4.40), and risk management (mean = 4.03). It is crucial for project managers to have a good understanding of cost management in order to effectively manage issues related to costly green materials and to prevent cost overruns. One of the interviewees pointed out that to ensure that costs stay within budget, cost estimates should be as close to the actual sum as possible during the project planning stage. Project managers have to quantify the costs and benefits of using green materials, despite market uncertainties, and assess the risk involved in their acquisition. One interviewee reported an example of this cost versus risk decision. The use of an innovative product could yield a high Green Mark score, but at the same time it posed a high risk of product failure because it lacked a proven track record. In fact, Table 6.6 shows that the decision-making skill is the most critical to effectively mitigate material and equipment-related challenges. Effective decision making also implies making the right choice within the shortest time to prevent delays.

6.3.2.6 *Knowledge Areas and Skills for Labor-Related Challenges*

As shown in Table 6.5, project managers need to understand human resource management (mean = 4.40) and communication management (mean = 4.00) in order to effectively address labor-related challenges. PMI also endorses human resource management as important for dealing with labor-related challenges (Project Management Insititute (PMI), 2013). Project managers have to provide appropriate and timely training for their workers and, in addition to other topics, some workers may need training on green construction. Communication management is vital to disseminating information and work procedures so that the workers can carry out their work effectively. The information must be correct and timely to prevent mistakes. This result is shown, along with the analysis on critical skills that can mitigate labor-related challenges, in Table 6.6.

6.3.2.7 *Knowledge Areas and Skills for External Challenges*

The respondents ranked knowledge of schedule and planning management as the most important to respond to external challenges (Table 6.5). With good planning knowledge, project managers may be able to allocate contingency time to buffer unforeseen situations and the lengthy approval process. It is notable that there are several knowledge areas with mean values below 3 in this category. This may imply that the external challenges posed to project managers would not be fully remediated by enhancing only the listed knowledge areas. The required skill for addressing external challenges was problem solving (Table 6.6), suggesting that such challenges may pose a great amount of uncertainty and thus project managers need to be equipped with the skills to manage unexpected events.

6.3.3 Knowledge Areas and Skills: Green vs Conventional Construction Projects

In the survey, respondents provided their assessment of knowledge areas and skills necessary for both green and conventional construction. The following sections present a comparison of their responses.

6.3.3.1 *Knowledge Areas*

As shown in Table 6.7, knowledge of schedule management and planning for conventional construction projects yielded the highest mean score of 4.77, followed by knowledge of health and safety management (mean = 4.73), and knowledge of human resource management (mean = 4.60). On the other hand, for green construction projects, knowledge of cost (mean = 4.90) and communication (mean = 4.80) management were considered more critical than schedule management and planning (mean = 4.77).

To further examine the rankings between conventional and green construction projects, the Spearman's rank correlation was calculated and its statistical

TABLE 6.7 Management Knowledge Area: Conventional vs Green Projects

	Conventional		Green			
Management Knowledge Area	*Mean*	*Rank*	*Mean*	*Rank*	**Rank correlation**	***P*-Value**
Cost management	4.57	5	4.90	1	.037	.920
Communication management	4.17	9	4.80	2		
Schedule management and planning	4.77	1	4.77	3		
Health and safety management	4.73	2	4.77	4		
Risk management	4.33	7	4.67	5		
Conflict and dispute management	4.23	8	4.43	6		
Stakeholder management	4.57	4	4.40	7		
Materials resources management	4.37	6	4.33	8		
Claims management	4.10	10	4.23	9		
Human resources management	4.60	3	4.23	10		

significance was tested. The rank correlation was found to be .037 with a P-value of .920, suggesting that there is no significant relationship in the ranking of the knowledge areas between the two groups. The reason for this disparity may be due to more emphasis being placed on other aspects of green construction projects. For instance, knowledge of cost management is important to green construction projects due to the higher cost of green materials, new systems and technologies, and specific equipment used for the projects. This result may prove that knowledge required for project manager competency varies by context, and green projects require different proficiencies than conventional construction.

6.3.3.2 Skills

As shown in Table 6.8, decision making, delegation, and analytical skills are ranked as the top 3 most important skills for green construction projects with a mean score of 4.87, 4.80, and 4.767, respectively. For conventional construction projects, the mean score for the decision-making skill is 4.57, delegation is 4.90 and analytical is 4.40. The Spearman's rank correlation (.675) indicates a fairly strong correlation between the two groups with a P-value of .032, which proves that the rank correlation is statistically significant. This means that the level of importance of the skills was perceived by the respondents to be similar for conventional and green projects.

6.4 SUMMARY

Based on a series of interviews, as well as a questionnaire survey, the top 10 critical challenges faced by project managers of green construction projects have been identified and those challenges are (1) the longer time required during the preconstruction process; (2) difficulty in the selection of subcontractors who provide green construction services; (3) uncertainty with green materials and equipment; (4) high cost of green materials and equipment; (5) increased meetings and coordination required with green consultants and engineers; (6) more frequent alterations and variations of the design during the construction process; (7) difficulty in comprehending the green specifications in the contract details; (8) unforeseen circumstances in executing green projects; (9) planning of non-traditional construction sequences; and (10) planning of different construction techniques.

Critical knowledge areas and skills that are essential to respond to the challenges were also identified. The most important knowledge areas are schedule management and planning, stakeholder management, communication management, cost management, and human resources management. In addition, the most important skills required to mitigate the challenges are analytical, decision-making, team working, delegation, and problem-solving skills.

TABLE 6.8 Management Skills: Conventional vs Green Projects

Management Skill	Conventional		Green		Rank Correlation	*P*-Value
	Mean	*Rank*	*Mean*	*Rank*		
Decision making	4.57	5	4.87	1	.675	.032
Delegation	4.90	1	4.80	2		
Analytical	4.40	6	4.77	3		
Team working	4.67	3	4.70	4		
Problem solving	4.57	4	4.67	5		
Leadership	4.83	2	4.60	6		
Negotiation	4.23	8	4.43	7		
Human behavior	4.03	10	4.37	8		
Chairing meetings	4.17	9	4.23	9		
Presentation	4.30	7	4.13	10		
Delegation	4.90	1	4.80	2		

Furthermore, a comparison of the knowledge areas and skills between conventional and green construction projects revealed that there are specific knowledge areas that should be strengthened in order to effectively manage green construction projects. This may be because more emphasis is placed on specific aspects of green building construction projects.

Chapter 7

Green Construction Project Financing: Policies, Practices, and Research Efforts

7.1 INTRODUCTION

It has been widely recognized that humanity is facing various grand challenges, such as climate change, resource depletion, and environmental degradation (Polzin, 2017; Li et al., 2013; Darko and Chan, 2016). Unfortunately, the building and construction industry is responsible for the rise of these challenges to a large extent, particularly because of its enormous consumption of energy (Olubunmi et al., 2016). According to the United Nations Environment Program (2001), the building and construction industry has become a big energy consumer that uses 40% of global raw materials and 40%–50% of global energy. Meanwhile, it has been proved to be a principal waste contributor, accounting for 40% of solid waste and greenhouse gas emissions worldwide (Hwang et al., 2017d; Chua, 2015). To address these critical issues, over the past two decades, the construction authorities and industries worldwide have been continuously advocating and promoting green construction, which is the innovative practice of creating resource-efficient and environmentally friendly structures (Olubunmi et al., 2016; Zuo and Zhao, 2014).

Continued transformation toward a truly sustainable built environment requires an all-hands-on-deck approach (Hoffman and Henn, 2008; Darko et al., 2017). Beyond innovative building design and materials (Chang et al., 2011; Rajendran et al., 2009), elevated resource efficiency standards (Wu and Low, 2010; Love et al., 2012), and comprehensive construction methods (Wu and Issa Raja, 2015; Ding et al., 2016), green construction project financing plays a critical role in fostering the development of a sustainable built environment (Marcelo and Daniel, 2013; Von Paumgartten, 2003). Green construction project financing is a broad term that refers to raising capital investments for construction projects or for companies that are dedicated to support the development of a low-carbon and more sustainable built environment (Hong Kong Financial Services Development Council, 2016). It is a strategic approach to incorporate various financial institutions (e.g., banks, insurance companies, property companies, investment trusts, institutional investors, and government)

Performance and Improvement of Green Construction Projects. https://doi.org/10.1016/B978-0-12-815483-0.00007-7

to direct their financial resources to the development of a sustainable built environment (Soundarrajan and Vivek, 2016). Green construction project financing is gaining attention currently. Particularly, momentum around green construction project financing has increased significantly due to the 2016 Group of Twenty (G20) Hangzhou Summit, in which the G20 Action Plan on the 2030 Agenda for Sustainable Development was released and the investment in green construction was strengthened (Group Twenty, 2016).

Currently, there is ample literature on green construction (Haapio and Viitaniemi, 2008; Olubunmi et al., 2016; Zuo and Zhao, 2014; Darko and Chan, 2017). However, the majority of the literature concentrates on the environmental benefits of green construction projects and the innovative design and construction technologies applied in green construction projects (Wong and Zhou, 2015; Ghaffarianhoseini et al., 2013), while a systematic review of green construction project financing remains lacking. Therefore, this chapter aims to review the ongoing policies, practices, and research efforts in the area of green construction project financing, and to identify the potential opportunities for future research actions. This chapter contributes to the body of knowledge by reviewing the state of the art of green construction project financing and revealing the potential areas for future research. Meanwhile, this chapter will be helpful to industry practitioners as it will summarize and present the lessons learned from current green construction project financing practices, thereby enhancing knowledge and skills in this area.

7.2 METHODS

A systematic literature review of green construction project financing was conducted. Beyond looking at the research findings obtained in academia, an assessment of the policies and practices implemented by the authorities and industry was undertaken. The literature reviewed in this chapter contains three sections, the policies and practices of green construction project financing implemented by representative developed economies, the efforts and initiatives launched by selected international organizations, and research outcomes presented in peer-reviewed journals and books.

Compared to undeveloped economies, developed economies are more active in green construction. This is because developed economies normally have better economic conditions to address sustainable built environment development (Olubunmi et al., 2016). Therefore, four representative developed economies were selected for review of their policies and practices adopted for green construction. These four developed economies are the United States (US), the United Kingdom (UK), Singapore, and Australia, all of which are active players in promoting green construction globally (Zuo and Zhao, 2014). In addition to developed economies, some international organizations are keen on promoting green construction as well (Haapio and Viitaniemi, 2008). They have launched various guidelines and initiatives to finance the development of a sustainable built environment over the past two decades (Olubunmi et al., 2016).

Therefore, this chapter also selected three notable and typical international organizations, namely, the United Nations, The Organization for Economic Co-operation and Development (OECD), and International Finance Corporation, to review their efforts and initiatives in green construction.

The research findings published in peer-reviewed journals and books from 2008 to 2017 has been reviewed to understand the research on green construction project financing. Web of Science is a powerful, widely-recognized academic search engine used by researchers worldwide to track the latest research progress in various areas (Zhao, 2017; He et al., 2017; Le et al., 2014). Thus, Web of Science was selected to track down those published journal articles relating to sustainable construction project financing. Four concise codes, including "TITLE: (green) AND TITLE: (finance)," "TITLE: (green) AND TITLE: (financial)," "TITLE: (sustainable) AND TITLE: (finance)," and "TITLE: (sustainable) AND TITLE: (financial)," were searched in Web of Science within the collections of Science Citation Index Expanded, Social Sciences Citation Index, and Emerging Sources Citation Index. Initially, a total of 148 papers were identified from the search. Then a careful visual examination was made to check the relevance of the identified papers. The visual examination results showed that the majority of the identified papers were irrelevant to the area of green construction and only 11 identified papers examined green construction project financing specifically. Thus, these 11 papers were selected for further review. Table 7.1 presents the details of these papers. In addition to peer-reviewed journals, an increasing number of books that discuss green construction have been published over the past decade (Finnegan, 2017). This makes books also an important source of literature for this chapter. Hence, an exhaustive search was conducted in Google Books to try to identify books related to green construction project financing. Finally, seven books that investigate green construction project financing were identified. The details of these seven books are summarized in Table 7.2.

The following sections review the green construction project financing practices implemented by selected developed economies, the green construction project financing initiatives launched by selected international organizations, and the research outcomes relating to green construction project financing published in peer-reviewed journals and books.

7.3 GREEN CONSTRUCTION PROJECT FINANCING PRACTICES AROUND THE WORLD

7.3.1 Green Construction Project Financing Practices in Developed Economies

7.3.1.1 The United Kingdom

To promote investment in sustainable built environment development, various policies and practices relating to green construction project financing have been developed by the United Kingdom. (Green Building Council, 2017).

TABLE 7.1 Journal Articles Identified From Literature Search

No.	Reference	Journal	Topic
1.	Zhan and de Jong (2017)	Sustainability	Financial vehicles for green construction projects
2.	Yamaguchi and Managi (2017)	Journal of Environment and Development	Financial vehicles for green construction projects
3.	Siamanta (2017)	Journal of Political Ecology	Review of green construction project financing market
4.	Streimikiene (2016)	Renewable & Sustainable Energy Reviews	Review of green construction project financing market
5.	Soundarrajan and Vivek (2016)	Agricultural Economics	Financial vehicles for green construction projects
6.	Fu et al. (2016)	Sustainability	Review of green construction project financing market
7.	Busch et al. (2016)	Business & Society	Review of green construction project financing market
8.	Lee et al. (2013b)	The Scientific World Journal	A financing model to overcome financial barriers for green construction projects
9.	Bose et al. (2012)	Energy Policy	A proposed solution for financing green construction projects
10.	Patlitzianas and Christos (2011)	Renewable & Sustainable Energy Reviews	A financing mechanism for sustainable development
11.	Lutzkendorf et al. (2011)	Building Research and Information	Review of financial stakeholders for green construction projects

These policies and practices have different forms, such as direct subsidy and creating institutions and mechanisms to catalyze investments in green projects. In 2012, The United Kingdom set up the world's first Green Investment Bank, the primary responsibility of which is to support and overcome barriers to investment in green projects (Green Investment Bank, 2017). Particularly, financing green construction projects and building retrofit projects are the Green Investment Bank's priority areas (Halper, 2015). Also, to accelerate investment in green retrofits to existing buildings, the United Kingdom launched a feed-in tariff in April 2010, under which private businesses and

TABLE 7.2 Books Identified from Literature Search

No.	Book Title	Author	Publisher
1.	New Financial Strategies for Sustainable Buildings: Practical Guidance for Built Environment Professionals	Finnegan (2017)	Taylor & Francis
2.	Delivering Sustainable Buildings: An Industry Insider's View	Malina (2012)	John Wiley & Sons
3.	Sustainable Finance and Banking: The Financial Sector and the Future of the Planet	Jeucken (2010)	Routledge
4.	The Law of Green Buildings: Regulatory and Legal Issues in Design, Construction, Operations, and Financing	Howe et al. (2010)	American Bar Association
5.	The Green Building Revolution	Yudelson and Fedrizzi (2010)	Island Press
6.	Global Trends in Real Estate Finance	Newell and Sieracki (2009)	John Wiley & Sons
7.	Green Building and Sustainable Development: The Practical Legal Guide	Furr et al. (2009)	American Bar Association

households would be paid for installing green features that can generate electricity from renewable sources such as wind, solar, and small-scale combined heat and power (Strielkowski et al., 2017). In addition to the feed-in tariff, the United Kingdom also launched another similar scheme named the Renewable Heat Incentive in November 2011 to encourage the installation of equipment that can generate renewable heat (e.g., solar thermal panels, renewably powered district heating, heat pumps, biomass boilers and deep geothermal systems) (Abu-Bakar et al., 2013). Another scheme of green construction project financing implemented in the United Kingdom is The Green Deal, which was launched in 2012 (Malina, 2012). The scheme was designed based on a "pay as you save" model: householders receive up-front finance in the form of a loan to pay for the installation of energy saving measures in their buildings and then pay back the loan using the savings obtained from their utility bills (Vaughan, 2012). However, in July 2015, this scheme was scrapped by the conservative government on the basis that it had failed to deliver its objectives (BBC, 2015).

7.3.1.2 The United States

Over recent decades, the United States government has provided a variety of financing options at different levels, including federal, state, and municipal levels, to support the development of a sustainable built environment (Tobias and Malachite, 2008). For instance, at the federal level, the US government initiated the Energy Efficiency and Conservation Block Grant program and the Weatherization Assistance Program to support units of local government, Indian tribes, and low-income families in implementing green retrofits to their buildings (Office of Energy Efficiency & Renewable Energy, 2017a,b). Also, the United States Department of Housing and Urban Development launched several financing programs, such as the Energy Performance Contracting and the Capital Fund Financing Program, to assist the development of housing projects that meet energy star construction and appliance purchase standards (Tobias and Malachite, 2008). Likewise, at the state level, a large number of grant, loan, leasing, and rebate programs have been started to finance sustainable development (Howe et al., 2010). As for the municipal level, various green bonds, such as Qualified Green Building and Sustainable Design Project bonds, Qualified Energy Conservation Bonds, and Clean Renewable Energy Bonds, were launched as the main vehicles to finance sustainable development (Howe et al., 2010). Additionally, the US federal government has enacted a large number of tax incentives to encourage the development and retrofit of housing to promote energy efficiency. These incentives include the Tax Credit for Home Builders, Tax Credits for Manufacturers of Energy Efficient Appliances, Tax Credits for Non-Solar Home Improvements, and Tax Credits for Home Solar Improvements (Tobias and Malachite, 2008). Moreover, the US Department of Energy funded a platform named the Database of State Incentives for Renewables and Efficiency, which contains 1324 cost reduction programs available throughout the United States. The primary mission of this platform is to provide more affordable options for green retrofitting projects in the United States to increase their sustainable improvements (Department of Energy, 2017).

7.3.1.3 Singapore

Singapore is a global leader in the area of sustainable built environment development. Over the past decade, the number of green buildings in Singapore has increased dramatically, from 17 in 2005 to over 2100 in 2014 (Hwang et al., 2017b). It has to be mentioned that financing programs from the Singapore government contributed significantly to achieving this progress (Hwang et al., 2017a). Currently, there are at least ten government financing programs for green construction being implemented in Singapore. The missions of these programs vary significantly, from providing credit facilities for building owners to carry out energy efficiency retrofits (e.g., Building Retrofit Energy Efficiency Financing), co-funding building owners in undertaking energy efficiency retrofits (e.g., Green Mark Incentive Scheme for Existing Buildings), through to supporting the private sector in developing green buildings (e.g., Green Mark Gross Floor Area Incentive Area) (Tay, 2015). These programs reflect real commitment from the local policymakers toward the development of a sustainable built environment in Singapore.

7.3.1.4 Australia

In 2008, the Australia Federal Government established a Green Building Fund amounting to AUD 90 million (approximately USD 71.24 million) to directly fund the green retrofit of existing buildings. This fund represents the first large-scale effort on the part of the Federal Government to fund the development of a sustainable built environment (Green Building Council Australia, 2008). Meanwhile, financing actions also exist at the state level. In 2015, South Australia introduced a green building upgrade and retrofit financing scheme named Environmental Upgrade Finance (Chua, 2015). This scheme enables property owners, occupants, service providers, and property managers to access a particular type of loan, which can be used to retrofit their buildings and is ultimately repaid to financers via a government charge levied on the property. This scheme can effectively address the barrier of "split-incentive" for the green retrofit to leased properties, which means the tenants enjoy the benefits of retrofitting but the owners bear the cost. This scheme makes good economic sense for property owners, occupants, and the business sectors that can provide clean technologies and solutions. (Chua, 2015; Sustainable Melbourne Fund, 2013).

7.3.2 Efforts and Initiatives From International Organizations

7.3.2.1 United Nations

In 1972 in the United Nations Conference on the Human Environment, Stockholm, Sweden, the United Nations established the United Nations Environment Program (UNEP) as its environmental conscience (Hill, 1975). Since its inception, UNEP has had a mandate to encourage economic growth compatible with the protection of the environment. In 1991, the United Nations launched the United Nations Environment Program-Finance Initiative (UNEP-FI), a partnership between UNEP and over 200 leading global banks (for example Deutsche Bank and HSBC Holdings), investment funds, and insurance companies (e.g., General Accident and National Provident) (UNEP Finance Initiative, 2017). The primary objective of UNEP-FI is to engage these leading financial institutions in a constructive dialogue about the nexus between environmental protection, economic development, and sustainable development (Dodds et al., 2012). A secondary objective of UNEP-FI is to catalyze private sector investment in green technologies and services, particularly those used for green construction. Additionally, UNEP-FI has also launched a considerable number of publications that discuss the framework and decision-making for green construction investments (Cherneva, 2012).

7.3.2.2 The Organization for Economic Co-operation and Development (OECD)

In 2016, the OECD established a Centre on Green Finance and Investment, whose primary mission is to help catalyze the transition toward a sustainable, low-emissions and climate-resilient economy through the development of effective institutions, policies, and instruments (OECD, 2016b). This Centre leverages the OECD's economics and policy expertise and serves as a global platform to engage with key financial institutions to achieve sustainable development. Also, the Centre has been trying to make itself a knowledge exchange center for the various players in green construction, such as government, the private sector, regulatory institutions, and academic and civil society (OECD, 2016b). Moreover, OECD has recently released a series of reports reviewing a variety of market mechanisms and policies regarding green construction project financing (OECD, 2017) and clean energy investment policy (OECD, 2016c). Furthermore, OECD has launched several reports that summarize the rationales, mandates, and financing activities of green investment banks publicly capitalized (OECD, 2016a), the barriers to attracting international investment in sustainable development (OECD, 2015b), and the various channels that can mobilize institutional investment in sustainable development (OECD, 2015a). All these reports have provided practical information and evidence-based analysis to industry practitioners and policy makers, improving their understanding of financing for green construction.

7.3.2.3 International Finance Corporation

International Finance Corporation (IFC), a member of the World Bank Group, is the largest global development institution that focuses exclusively on development in developing countries (IFC, 2017a). One of IFC's primary objectives is to help the private sector achieve sustainable development through innovative investments and financing. Since 2005, IFC has invested about $15.3 billion in long-term financing for renewable power, energy efficiency, and green buildings (IFC, 2017a). Additionally, IFC was one of the earliest issuers of green bonds. In 2010, IFC launched the green bond program to help catalyze the market and unlock investment for green construction projects in the private sector (IFC, 2017b). As of 2014, IFC has supported 93 green bond eligible projects worldwide amounting to close to USD 3 billion (IFC, 2016). Moreover, IFC plays a leadership role in developing principles, guidelines, and procedures for the green bond market, providing the issuers and investors practical guidance in launching a credible green bond for green construction projects (IFC, 2017b). In addition, IFC has also released a green bond impact report annually, which has been widely referenced by numerous financial sectors and institutions around the world.

7.4 RESEARCH FINDINGS WITHIN GREEN CONSTRUCTION PROJECT FINANCING

Based on a systematic review of the identified journal articles and books, it was found that current research on green construction project financing has concentrated on four areas, namely, review of financial stakeholders and the market for green construction, benefits and barriers of green construction project financing, the identification of financial vehicles for green construction, and innovative models and mechanisms for green construction project financing. In the following sections, each area will be briefly introduced and described. Insights into, and reasons for, current practices, are also discussed.

7.4.1 Review of Financial Stakeholders and Market of Green Construction

Lutzkendorf et al. (2011) listed seven major financial stakeholders engaged in green construction projects: (1) the financial sector, such as banks and insurance companies, (2) project developers, (3) property companies and housing companies, (4) initiators, providers and managers of property funds and Real Estate Investment Trusts, (5) Institutional investors, such as investment companies, investment funds, and pension funds, (6) individual/private investors, and (7) government. However, Lutzkendorf et al. (2011) pointed out that, although most financial stakeholders (particularly banks, insurance companies and investors) have already been engaged with the development of a sustainable built environment, their understanding of these types of projects remains limited. Similar conclusions were also obtained by Busch et al. (2016) who pointed out that current financial stakeholders have not been ready to embrace green construction finance. To address this critical issue, both Lutzkendorf et al. (2011) and Busch et al. (2016) suggested carrying out education and training for the employees of these financial stakeholders to increase their knowledge and competencies in green construction finance.

In addition to financial stakeholders, the green construction project financing market was also investigated. Fu et al. (2016) discovered that the Private Participation in Infrastructure investments had a significant promotional effect on the development of financial markets in terms of size and liquidity using data from 33 developing countries during the period 1997–2012. They also confirmed the significant mediator effect of financial market size for the positive relationship between Private Participation in Infrastructure investments and financial market liquidity. By contrast, Siamanta (2017) found some side effects of green construction finance caused by overinvesting. Siamanta (2017) examined the growth of photovoltaic projects in post-crisis Greece, and found that this country had overemphasized green development, resulting in a few green economy entrepreneurs monopolizing the whole country's market. This led to the accumulation of capital by a few large renewable energy resources companies and imposed significant negative impacts on the livelihoods of small business electricity consumers and small/medium-sized photovoltaic investors.

7.4.2 Green Construction Project Financing: Benefits and Barriers

Several benefits of implementing green construction project financing were identified from the literature. First, implementing green construction project financing helps promote the diffusion of green construction technologies in the construction industry. Meanwhile, it can add value to those businesses, organizations, and corporations that engage in green construction financing by creating for them a positive social image and bringing them a further green edge (Soundarrajan and Vivek, 2016). Moreover, implementing green construction project financing can enable financial institutions to face less risk compared to financing traditional construction projects. This is because the building owners and contractors who embrace green technologies are likely to be more risk management-minded and practice greater care in building construction, operation, and maintenance (UNEP Finance Initiative, 2010).

In addition to these benefits, several barriers that hinder the involvement of financial institutions in green construction were uncovered. These barriers are split incentives, lack of consistent, accurate data analysis and interpretation, lack of knowledge about green construction, high upfront costs, and regulatory gaps. The barrier of split incentives is actually not unique to financial institutions but rather is a generic barrier to anyone getting involved in green construction. It is a principal-agent problem basically (UNEP Finance Initiative, 2010). Particularly, this barrier is more prominent in rental buildings as it makes building owners bear the cost of improving energy efficiency while the tenants enjoy the savings in utilities. Because of this barrier, a large number of owners of rental buildings are reluctant to invest in green retrofitting of their buildings (Chua, 2015). The relative lack of consistent, accurate data analysis and interpretation is another significant obstacle to financing green construction. This is because most financial institutions will not approve investment in green construction projects unless attractive financial returns and benefits from future projects can be proved conclusively (Weber et al., 2016). However, this is not yet feasible on account of the many criteria, approaches, and standards being used for certifying sustainable built structures, and, consequently, data cannot be analyzed and explained using a unified standard (UNEP Finance Initiative, 2010). Lack of knowledge about green construction is also a major barrier for the implementation of green construction project financing. While demand for, and understanding of, sustainable and green buildings have increased markedly over the past several years, the knowledge of many professionals in the financial institutions has not kept pace. A large number of professionals in these financial institutions do not have a sound understanding of green construction practices, making them unable to deal effectively with the financing for green construction projects (UNEP Finance Initiative, 2010). High upfront costs have also been considered a significant barrier. According to the California Sustainability Alliance (2009) and Lee et al. (2013b), high upfront cost is a particular barrier to investors engaging in green leasing. Another barrier to the expansion of green construction project financing lies in regulatory gaps, which refers to the absence of adequate laws (including regulations and monitoring and enforcement systems) to govern the implementation of green construction finance (Soundarrajan and Vivek, 2016).

7.4.3 Identification of Financial Vehicles for Green Construction

Five prevailing financial vehicles that have been widely used for green construction were identified, namely, bank loans, green bonds, international assistance programs, government grants and tax incentives, and private capital.

7.4.3.1 Bank Loans

Bank loans are one of the most common financial vehicles for green construction due to the flexibility in terms (e.g., long-term and short-term) and of type (e.g., secured, unsecured, demand, subsidized, and concessional loans) (Zhan and de Jong, 2017). Over the past two decades, numerous banks around the world have been arranging loans to finance the development of a sustainable built environment. For instance, Citigroup launched a green lending initiative amounting to USD 50 billion in May 2007 (Tobias and Malachite, 2008). Bank of America unveiled a USD 20 billion, 10-year green mortgage program in March 2007 (Tobias and Malachite, 2008). Through 2009, Wells Fargo had loaned more than USD 1.7 billion to approximately 35 LEED projects (Furr et al., 2009). However, according to the review, despite their active involvement, banks' attitudes toward loans for sustainable development seem divided according to the size of the bank. Furr et al. (2009) noticed that, compared to larger banks, the smaller regional banks are more supportive of financing sustainable development. They offer various preference programs for sustainable development, such as preferred project review, reduced interest rates, exemption for initial premium of mortgage for green commercial building development, and increased loan-to-value ratios. By contrast, the larger banks are relatively more conservative. They normally treat sustainable development projects the same as conventional development projects. They do take green specific considerations into account when doing project reviews. But as a whole, they still grant loans based on the fundamentals underlying the projects. In addition to the merits of bank loans, some of their disadvantages were also reported in the literature. These disadvantages included high costs, complicated procedures to obtain approval from banks, and limited credit amounts (Zhan and de Jong, 2017).

7.4.3.2 Green Bonds

According to The World Bank (2015), a bond is a form of debt security that can be bought and sold between parties, and that a green bond is a debt security that is issued to raise capital specifically to support environmental projects, including green construction projects. According to the review, in the past decade, green bonds have received increased attention from financial institutions and institutional investors and have become popular in financing green construction projects. Especially, multilateral banks are the most active financial institutions in issuing green bonds to support green construction (The World Bank, 2015). For instance, The African Development Bank issued a USD 500 million green bond in October 2013 to support the development of a sustainable built environment in Africa (The World Bank, 2015). In 2007, The European Investment Bank issued a EUR 600 million green bond that mainly focused on renewable energy and energy efficiency retrofits to buildings (European Investment Bank, 2007). Since 2008, The World Bank has issued over USD 10 billion equivalent in green bonds through more than 130 transactions in 18 currencies to support at least 70 sustainable development projects in the developing countries around the world (World Bank Treasury, 2017; The World Bank, 2015). The green bonds issued by these multilateral banks have catalyzed the financial market for green construction projects tremendously and helped attract substantial investment in these types of projects. Furthermore, the review also shows that the private sector has recently become interested in issuing green bonds to finance green construction projects. A typical case is the Regency Centers Corporation, a real estate investment trust that issued USD 250 million of green bonds in May 2014 to finance the construction of shopping malls that meet the standards of the US Green Building Council (Regency Centers Corp., 2014).

7.4.3.3 International Assistance Programs

Recently, several international financing institutions, such as The Global Environment Facility (GEF) Trust Fund, Special Climate Change Fund, Least Developed Countries Fund, Nagoya Protocol Implementation Fund, and Adaption Fund, have provided a series of international assistance programs to support sustainable development around the world, particularly in developing countries (Global Environment Facility, 2017). One of the priorities of these international assistance programs is to finance sustainable built environment development (Global Environment Facility, 2016). For instance, The Global Environment Facility Trust Fund offered a USD 6.16 million grant to sponsor the development of Sino-Singapore Tianjin Eco-City. The grant was allocated to four areas, namely, integrated water technical assistance, energy utilization, transport management, and construction of green buildings (The World Bank, 2017). A commonality of the international assistance programs is that the amounts involved are limited. Nevertheless, these international assistance programs are still helpful and even play vital roles in financing sustainable development. This is because international assistance programs are actually walking advertisements for green construction projects and can help attract more capital and investment. The reason behind this is simple. Once a project is sponsored by an international assistance program, its feasibility and viability has been proved because the assessment processes of these international assistance programs are very strict and rigorous in most cases.

7.4.3.4 Government Grants and Tax Incentives

In order to promote green construction, governments around the world also provide an increasing amount of support in terms of grant preferences. For example, the United States Department of Housing and Urban Development has launched a large number of financing programs, such as Energy Efficiency and Conservation Block Program, State Energy Program, and Public Housing Capital-Energy Conservation Retrofit Investment Program to assist housing projects and help them meet energy star construction and appliance purchase standards (Tobias and Malachite, 2008; Howe et al., 2010). Singapore initiated a series of incentive plans (e.g., Green Mark Incentive Scheme for New Buildings in 2006 and Green Mark Incentive Scheme for Existing Buildings in 2009) to encourage local developers, building owners, and project consultants to develop and construct green buildings (Hwang et al., 2017a,b). Some countries' governments are relatively more straightforward. They appropriate cash subsidies to green construction projects directly. For example, from 2013 to 2015, the city government of Tianjin, China appropriated CNY 299.72 million (approximately USD 44.18 million) to Sino-Singapore Tianjin Eco-City to support the development of this green construction megaproject (Zhan and de Jong, 2017). In addition to government grants, tax incentives enacted by governments around the world are another financial vehicle to encourage the development of green construction (Tobias and Malachite, 2008). In the United States, numerous tax incentives have been launched, such as the Energy Investment Tax Credit, the Tax Credit for Energy Efficiency Improvements to Existing Homes, Residential Energy-Efficient Property Credits, the Energy-Efficient Commercial Buildings Deduction, and the Energy-Efficient New Homes Tax Credit (Howe et al., 2010). During the development of Sino-Singapore Tianjin Eco-City, it has been reported that the Tianjin government also refunded part of the taxes derived from the project in order to encourage its development (Zhan and de Jong, 2017).

7.4.3.5 Private Capital

Green construction requires long-term investment and sustained financing (Chan et al., 2009; Fabian, 2015). Traditionally, public budgets are an important financial source for green construction. But given the strains on public budgets, private capital is needed for green construction as well (Love et al., 2015a). Over the past 10 years, an increasing amount of private capital from private companies has flowed into the green construction sector (Yudelson and Fedrizzi, 2010; Furr et al., 2009). In the beginning, private companies, particularly those real estate private companies, invested in a sustainable built environment mainly to create a positive image in terms of corporate social responsibility. However, gradually, they realized that such investments can bring them reasonable economic returns. Therefore, they increased these investments (Yudelson and Fedrizzi, 2010). Recently, a large number of Real Estate Investment Trusts around the world have invested in green buildings. For example, Corporate Office Properties Trust and Liberty Property Trust developed LEED Silver buildings for corporate tenants (Liberty Property Trust, 2014, Hamilton, 2007). They said that green buildings are wiser investments because they are leasing up faster, are cheaper to operate, and are able to attract tenants of better quality (Yudelson and Fedrizzi, 2010). Another case is of the Corporate Office Properties Trust. One of its green construction projects at 318 Sentinel Drive, in the National Business Park in Annapolis Junction, Maryland, received the 2005 National Associations of Industrial and Office Properties Green Development Award. The project, a four-story, 125,000-square-foot office building, was fully leased even before the completion of construction (Hamilton, 2007). In addition to investing to earn benefits, numerous investors have also donated to nonprofits and private colleges and universities to build new green buildings or renovate old buildings with green features. For example, the Ecotrust organization in Portland, Oregon, received a major gift from a single donor to renovate a 100-year-old warehouse into a 70,000-square-foot modern building. This project won the second LEED gold-certified project in the United States when it opened in 2001 (Yudelson and Fedrizzi, 2010).

7.4.4 Innovative Models and Mechanisms for Green Construction Finance

Another area of green construction project financing research explores innovative financing models and mechanisms. Realizing that the development of a sustainable built environment may face barriers in attracting sufficient funds, Lee et al. (2013b) suggested a financing model that includes a government guarantee. Similarly, Yamaguchi and Managi (2017) introduced a Green Net National Product-Linked Bond to finance green construction. This bond would lead the nation's wealth to sustainable development, raise future tax revenue if combined with tax reform, realize intergenerational sharing of well-being with current investors, and be a good opportunity to diversify assets to lower investment risk and volatility. Patlitzianas and Christos (2011) introduced a modern financial mechanism named "JESSICA" to finance sustainable urban development projects. Under this mechanism, various forms of funds such as equity, loans, and guarantees were provided to fund the development of sustainable urban projects, and the returns, for instance loan repayments, would be used to reinvest in new urban development projects and thereby bolster sustainability. In order to improve the financing performance of green construction projects, it is important and necessary to examine the relationship between the nature and the sources of financial flows. Hence, Bose et al. (2012) developed an approach named "financial gradients" to analyze financial flows in green construction projects. This approach can help identify and secure a single, long-term, stable inflow of finance to sustain the process of green construction projects.

7.5 FUTURE RESEARCH DIRECTIONS

Based on the systematic review above, four future research directions in green construction finance can be deduced, including the identification of financial issues in green construction projects, the examination of critical drivers for implementing green construction finance, the investigation of the performance of diverse financial vehicles for green construction projects, and the development of best practices for implementing green construction project financing.

7.5.1 Identification of Financial Issues in Green Construction Projects

One important point that has been omitted in the current literature is the need to identify the financial issues in current green construction projects. Currently, financial issues in traditional construction projects have been adequately addressed (De Marco and Mangano, 2013; Chiang and Cheng Eddie, 2009; Taylor and Hensey, 1990; Badu et al., 2013). Various financial issues in traditional construction projects, such as inflation and fluctuation, shortages of investment capital, poor credit ability, interest rate, liquidity, and currency exchange restrictions, have been comprehensively investigated (Lam and Chow, 1999; Xenidis and Angelides, 2005; Aladağ and Işik, 2017; Wibowo and Kochendörfer, 2005). However, similar research efforts have not been made in green construction projects. Therefore, it is necessary and important to bridge this knowledge gap by identifying and summarizing the prevailing financial issues in current green construction projects. By identifying these financial issues, authorities and industry can be provided with a full view of the obstacles hindering green construction project financing and can then develop effective improvement strategies.

7.5.2 Examination of Critical Drivers for Implementing Green Construction Finance

Current literature has addressed the benefits of, and barriers to, implementing green construction project financing (Soundarrajan and Vivek, 2016; Chua, 2015; Weber et al., 2016; UNEP Finance Initiative, 2010). However, it has not addressed the drivers for implementing green construction project financing. Compared to examining the benefits and barriers, examining drivers is relatively more important because examining drivers means looking for the critical factors that can promote green construction project financing. Therefore, one of the most important research directions can be examining the drivers for implementing green construction project financing. Moreover, if the contributions of different drivers can be quantified or visualized with the aid of empirical studies, then the most important drivers can be identified, which can help develop tailored strategies to drive green construction project financing further.

7.5.3 Investigation of Financial Vehicles for Green Construction Projects

Although various financial vehicles, such as bank loans, green bonds, international assistance programs, government grants and tax incentives, and private capital have been identified in the current literature (Zhan and de Jong, 2017; Tobias and Malachite, 2008; Furr et al., 2009; Howe et al., 2010; The World Bank, 2017), they are yet to be investigated systematically. Thus, the investigation of financial vehicles for green construction projects is another promising avenue of research. Such an investigation might examine two aspects. First, there is a necessity to investigate the strengths and limitations of various financial vehicles. This kind of research effort would not only help building owners and developers to choose the most appropriate financial vehicle, or the best combination of different vehicles to fund green construction projects, but would also provide financial institutions with a good opportunity to upgrade their vehicles accordingly to make them more financially viable. Second, it is imperative to develop some systematic approaches to measure the financial performance of different financial vehicles, which can help building owners, developers, and financial institutions to lock down the best financial vehicles. Furthermore, it would also be interesting to compare the performance of financial vehicles in traditional and green construction projects, which would help the relevant stakeholders to decide whether they would like to have a traditional building or a sustainable building.

7.5.4 Development of Best Practices for Implementing Green Construction Finance

Existing literature reveals that current financial stakeholders are not yet ready to embrace the tide of green construction project financing due to their limited knowledge in green construction (Busch et al., 2016; Lutzkendorf et al., 2011). Thus, it is imperative to direct some research efforts to equip financial stakeholders with proper knowledge to make them capable of addressing green construction finance. Actually, it would be even more meaningful if some research effort could be applied to develop a set of best practices for implementing green construction project financing. Such a set of best practices can help financial stakeholders gain a deeper understanding of green construction finance, improve the skills of its workforce, and enable them to respond more quickly to new challenges and problems. The various financial stakeholders play different roles in green construction project financing, and thus development of best practices should be customized to the particular stakeholder in question.

7.6 SUMMARY

Investment in green construction will definitely take place on a far greater scale in the future due to the promotion of sustainable development. However, a systematic review of green construction project financing remains lacking. Therefore, a systematic review of practices, polices, initiatives, and research efforts in this area has been presented. Potential opportunities for future research actions were also discussed. First, this chapter reviewed the green construction project financing practices implemented in representative developed economies, such as the United Kingdom, the United States, Singapore, and Australia. Then, the efforts and initiatives in green construction launched by selected international organizations, such as the UN, OECD, and IFC, were reviewed. After that, the research efforts in green construction project financing published in peer-reviewed journals and books were reviewed. Based on the review, four major research themes within the area of green construction project financing were identified, namely, the review of financial stakeholders and the market for green construction, benefits and barriers to green construction project financing, identification of financial vehicles for green construction, and innovative models and mechanisms for green construction project financing. This chapter also identified four future directions for green construction project financing research, which are the identification of financial issues in green construction projects, the examination of critical drivers for implementing green construction project financing, the investigation of financial vehicles for green construction projects, and the development of best practices for implementing green construction project financing.

Although the objective of this chapter has been achieved, there are some limitations. First, because of the limited time and space, this chapter could not cover all the green construction finance practices and initiatives of every developed economy and international organization. Second, due to the limit of the search code, some studies and books that investigated green construction project financing partly in their contents might have been omitted.

In spite of the limitations, the findings from this chapter are still valuable. This chapter has presented the prevailing policies, practices, and research efforts of green construction project financing implemented worldwide. This can provide practitioners with a holistic view of green construction project financing, thereby enhancing their capabilities in this area. At the same time, this chapter identified significant future research opportunities in green construction project financing, which will be helpful and informative to academic researchers.

Chapter 8

Cost Premium, Performance and Improvement of Green Construction Projects

8.1 INTRODUCTION

Green building developments are greatly discouraged due to the perceived higher costs, commonly known as green cost premiums (Dodge Data & Analytics, 2016). Currently, there is no standardized definition for the green cost premium and no clear methodology to describe its components, nor any clear methodology to estimate its magnitude. (Houghton et al., 2009; Dwaikat and Ali, 2016). Kats (2010) defined the green cost premium as the differential cost between the green and conventional "version" of the same building. Houghton et al. (2009) proposed a definition for the green cost premium as the additional design and construction costs associated with specific green components. Moreover, in the report of Dodge Data & Analytics (2016), the higher initial costs, which were ranked as the top challenge in green building, were actually additional capital costs because they were used to make comparisons with the decreased operating costs of green buildings. In light of the above review, the green cost premium can be defined as the additional capital costs of green building features.

In spite of the benefits of green buildings and the various efforts made to promote a sustainable built environment, key building stakeholders are still somewhat skeptical about the financial benefits that green buildings can deliver. Many industry professionals have the perception that the design and construction costs of green buildings are 10%–20% higher than those of conventional buildings (WorldGBC, 2013). In light of this, the higher costs associated with "going green" are thought of as the most common reason that hinders the wide development of green buildings (Robichaud and Anantatmula, 2011; McGraw-Hill Construction, 2006b).

The design and construction costs are perceived as contributing to the green cost premium. Green building projects generally have more complex designs compared to traditional building projects (Johnson, 2000). In order to achieve sustainability, green building projects generally require the use of special specifications, materials, construction methods and building practices

Performance and Improvement of Green Construction Projects. https://doi.org/10.1016/B978-0-12-815483-0.00008-9

(Robichaud and Anantatmula, 2011; Lam et al., 2010). Moreover, productivity in the design and construction of green building projects is currently lower than that of traditional projects because practitioners still need time to learn and become proficient in these technologies. Furthermore, unfamiliarity with green technologies and technical difficulties during the construction process not only affect the project schedule, but also lead to cost increases by rework (Tagaza and Wilson, 2004; Hwang et al., 2009).

In properly facing this barrier, some countries, such as the United States, the United Kingdom, and Australia, investigated the high green cost premium (Dwaikat and Ali, 2016). Kats (2010) conducted a large-scale study based on extensive financial and technical analyses of >150 green buildings in the United States and 10 other countries. The results showed that green buildings cost roughly 2% more to build than traditional buildings. Moreover, Kim et al. (2014) concluded that the green cost premium for residential project development in Los Angeles was 10.77%. In addition, Houghton et al. (2009) found that the green cost premium for healthcare buildings in the United States ranged from 0% to 5% without any financial incentives. In the United Kingdom, the Building Research Establishment (BRE) and Cyril Sweett (2005) asserted that the green cost premium ranged from 0% to 7%. In Australia, Davis Langdon (2007) reported that the impact on the construction cost ranged from 3% to 5% for Five Star rating, and >5% for Six Star noniconic design solutions. Dodge Data & Analytics (2016) also conducted a study on the challenges to green building and identified the higher perceived upfront costs as one of the top three challenges in nearly all of the 13 surveyed countries. This challenge was selected by over 50% of the respondents only in the United States (70%), Mexico (54%), Colombia (67%), Germany (52%), the United Kingdom (52%), and China (60%).

However, compared with the other leading countries, Singapore lacks knowledge and data on green cost premiums. Furthermore, there is limited research on the green cost premium and cost performance in Singapore. Wong et al. (2003) examined the initial cost implication of having a green roof in Singapore and concluded that the initial costs were 82%, 36%, and 50% higher for an inaccessible extensive green roof, an accessible intensive roof with shrubs, and an accessible intensive green roof with trees, respectively, than those for equivalent conventional roofs. Deng and Wu (2014) investigated the economic returns to residential green building investment in Singapore from the developers' perspective to ameliorate public doubt on the financial viability of investments in energy efficiency. An investment in energy-efficient real estate development can only be financially sustainable if the "green price premium" is large enough compared to the additional "green costs." The results showed that the developers claimed 4% market premium of Green Mark rated units at the presale stage to cover the additional costs of energy efficiency during construction. Goh (2016) investigated the whole-life costs of nonresidential green-rated building developments in Singapore in order to propose a whole-life building cost index, with the only limitation that the formulation of the proposed index lacked data support.

Besides the green cost premium, cost overrun is another hindrance. Because green buildings use more technologies and material with less environmental impact, they are more complicated than conventional buildings. Moreover, many requirements to achieve a green certificate, and the shareholders' unfamiliarity of the requirements and technologies, tend to lead to cost overruns, project delays and productivity losses (Nalewaik and Venters, 2010; Construction Industry Institute (CII), 2008). The empirical study of Hwang and Leong (2013) also found that, in Singapore, about 32% of green construction projects were completed behind schedule, while 16% of conventional construction projects were delayed.

Cost performance indicates how well costs are kept under control, that is, over budget or under budget. Cost overrun is generally a symptom of poor management. Thus, the research on cost overrun assessment (Chandramohan et al., 2012), identification of important factors affecting cost performance (Son et al., 2015), and comparison of the impact of preproject planning between green and conventional construction projects on cost performance (Kang et al., 2013), has begun to draw attention from researchers. In addition, considering the unique characteristics of green buildings and green construction, Robichaud and Anantatmula (2011) tried to improve the chances of delivering the project within acceptable costs by suggesting some construction management adjustments to conventional project management practices. However, there is, generally, still a need to investigate the actual cost performance of green building projects.

The aims of this chapter are to investigate the cost premiums and cost performance of green construction projects, and to come up with a series of feasible solutions that can reduce cost premiums in such projects.

8.2 METHOD AND DATA COLLECTION

This chapter adopted the questionnaire survey method previously elaborated in Chapter 3. First, an extensive literature review from multiple sources, such as government websites and reports from private institutions and journal papers, was carried out to provide a better understanding of the current market situation of green building and the issues relating to the cost premiums and cost performance of green construction projects. Then a questionnaire was subsequently developed to: (1) capture the current perceptions of professionals on the cost premium and cost performance of green construction projects; (2) identify the significant reasons for the cost premium; and (3) gauge the effectiveness of the solutions proposed to reduce the green cost premium and improve cost performance. The collected data were analyzed by the Statistic Package for Social Science (SPSS) statistical software.

The questionnaire first provided the definition of the green cost premium, which was the premise of the survey. Subsequently, the questionnaire presented questions meant to profile the companies and respondents. Furthermore, the respondents were asked to indicate the cost premium of green construction projects by different project type and size. The respondents were also asked to rate the significance of the reasons for the difference in the cost premiums between the green and conventional construction projects by using a five-point scale (1 = strongly disagree; 2 = disagree; 3 = neutral; 4 = agree; and 5 = strongly agree). Afterward, the number of conventional and green construction projects with different cost performance was indicated. Finally, the effectiveness of the solutions intended to reduce the cost premium of green construction projects and improve their cost performance was rated by using the five-point scale (1 = least efficient; 2 = somewhat efficient; 3 = neutral; 4 = efficient; and 5 = most efficient). In addition, postsurvey interviews were carried out with two green building professionals who had at least 3 years of experience in the green construction industry, especially in green construction cost management, to validate the findings obtained from the survey.

The population of this survey consisted of all the professionals who were from the BCA Green Mark past award winners, members of the Singapore Institute of Surveyors and Valuers (SISV), and registered contractors and licensed builders in the BCA directory who had more than 3 years of experience in the green construction industry and specialized in green building cost performance. The survey questionnaires were randomly sent out to the professionals via email. Finally, 30 responses were received. Although the sample size was relatively small, statistical analysis could still be performed because the central limit theorem holds true when the sample size is no <30 according to the generally accepted rule (Ott and Longnecker, 2010). The profiles of the respondents, companies and projects are provided in Table 8.1.

The respondents consisted of project managers, quantity surveyors, and contractors, and the majority of the respondents (70%) had at least 2 years' experience in the green construction industry, which indicates that the collected cost information is reliable. The percentages of respondents from the architecture, quantity surveying and contractor firms were 7%, 17% and 76%, respectively.

A total of 242 and 121 traditional and green building projects were recorded from the survey, respectively. The percentages of the three types of projects in conventional and green construction projects are generally comparable. The number of office building projects recorded from the survey is notably less than the number of commercial and residential building projects.

8.3 ANALYSIS RESULTS AND DISCUSSIONS

8.3.1 Cost Premium: Perception vs Actual

The overall perception of the survey respondents of the cost premium for green construction projects is summarized in Fig. 8.1. As seen in Fig. 8.1, 43% of the respondents perceived the green cost premium to be 5%–10%, followed by 34% and 23% of the respondents who perceived it to be 10%–15% and 0%–5%, respectively. None of the respondents perceived the green cost premium to

TABLE 8.1 **Profiles of Respondents, Companies and Projects**

Profile	Number	Percentage (%)
Job title	8	27
Project manager	5	17
Quantity surveyor	17	56
Contractor		
Years of experience in green construction		
<1 year	9	30
1–2 years	0	0
2–3 years	5	17
3–4 years	6	20
>4 years	10	33
Type of company		
Architecture	2	7
Quantity surveying	5	17
Contractor	23	76
Projects		
Traditional projects		
Commercial	124	51
Offices	20	8
Residential	98	41
Green projects		
Commercial	57	47
Offices	10	8
Residential	54	45

be above 15%. This result was in line with the argument made by Houghton et al. (2009) that the green cost premium was getting lower due to decreasing capital cost over time. Furthermore, according to the report from WorldGBC (2013), building professionals both with experience and without any experience in green projects, tended to perceive the green cost premium to be up to 13% and 18%, respectively, which was not significantly different from the results of this survey.

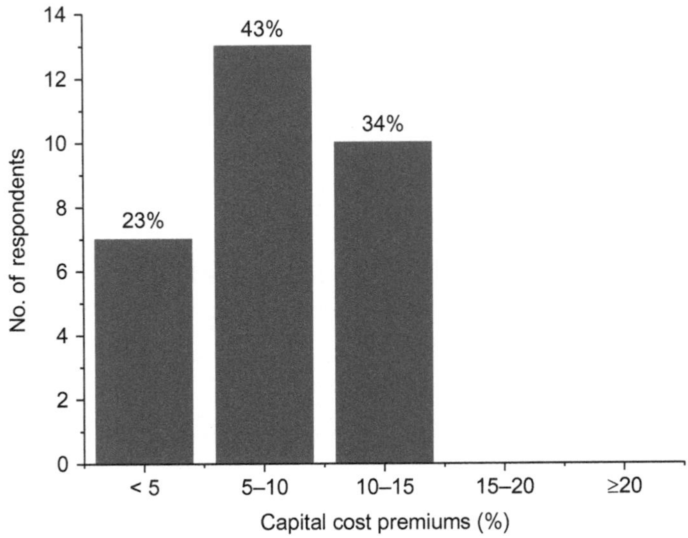

FIG. 8.1 Overall perception on green cost premiums. *(Green Building Construction Projects in Singapore: Cost Premiums and Cost Performance, Project Management Journal, 2017. Copyright and all rights reserved. Material from this publication has been reproduced with the permission of PMI.)*

In addition, this chapter examined the actual cost premium of green construction projects that were undertaken by the respondents. The cost premium for green projects by project size (less than S$5 million, S$5 million to less than S$50 million, S$50 million and above) and type (green commercial, office and residential buildings) is summarized in Table 8.2. The result indicated that there was indeed a cost premium for going green, generally ranging from 0% to <15%, regardless of the project type and size. This result was consistent with the overall perception of the green cost premium presented in the previous section.

8.3.1.1 Actual Cost Premiums By Project Type

To obtain the mean cost premium by project size and type, the mid-values of the four ranges of the premium, 2.5%, 7.5%, 12.5%, and 17.5%, were used. The mean cost premium of green construction projects by project size and type is shown in Table 8.3.

The overall mean of the green cost premium ranged from 2.5% to 12.5%. This result was comparable to the result from Kansal and Kadambari (2010), in which the conclusion was that "the initial cost of a green building is 7.5% more than the ordinary building." Additionally, it is clear that green residential has the highest cost premium, followed by green commercial and green office, for three different size classifications. The mean for green residential building was very close to the result from Kim et al. (2014), in which the conclusion was that "green residential building costs are 10.77% more than that of the

TABLE 8.2 **Cost Premium of Green Building Projects By Project Type and Size**

Project Size (S$ Million)	Capital Cost Premiums (CCP)	No. of Green Commercial	No. of Green Office	No. of Green Residential
<5 (small)	0% ≤ CCP < 5%	1	2	0
	5% ≤ CCP < 10%	9	1	1
	10% ≤ CCP < 15%	0	0	4
	15% ≤ CCP < 20%	0	0	0
	20% ≤ CCP	0	0	0
5–50 (medium)	0% ≤ CCP < 5%	12	0	0
	5% ≤ CCP < 10%	9	4	2
	10% ≤ CCP < 15%	9	0	3
	15% ≤ CCP < 20%	2	0	2
	20% ≤ CCP	0	0	0
50 and above (large)	0% ≤ CCP < 5%	11	3	32
	5% ≤ CCP < 10%	4	0	5
	10% ≤ CCP < 15%	0	0	5
	15% ≤ CCP < 20%	0	0	0
	20% ≤ CCP	0	0	0

traditional residential buildings." The relatively higher cost premium for green residential projects could be due to the respondents' lack of green expertise in green residential as compared to green commercial and office building projects. Respondents without sufficient green building expertise in residential projects would have difficulty in complying with the green specifications, hence, leading to delays and increased costs (Architecture Week, 2001).

TABLE 8.3 Mean Cost Premium of Green Construction Projects By Project Size and Type

	Mean of Green Cost Premiums		
Project Size (S$ Million)	*Commercial (%)*	*Office (%)*	*Residential (%)*
<5 (small)	7.0	4.2	11.5
5–50 (medium)	7.7	7.5	12.5
50 and above (large)	3.8	2.5	4.3

TABLE 8.4 ANOVA and TPH Results By Project Type

Project Size (S$ Million)	*P*-Value (ANOVA)	*P*-Value (Tukey Post Hoc)	
<5 (small)	.000	C vs R	.002
		R vs O	.000
		C vs O	.110
5–50 (medium)	.045	C vs R	.038
		R vs O	.197
		C vs O	.998
50 and above (large)	.601	No difference	

Note: C, O, and R denote commercial, office, and residential building, respectively.

Since the BCA Green Mark for Office Interior was launched in 2009, learnings in the area of green offices have increased, resulting in the lowest green cost premiums (Nalewaik and Venters, 2010). The opinions expressed in post survey interviews revealed that it may be much easier to comply with the green specifications for green office projects. Another reason could be due to the relatively small data set for green office building projects recorded from the survey responses.

A one-way analysis of variation (ANOVA) test was performed to test whether the project type has a significant effect on the green cost premiums. The null hypothesis H_0 is that there is no statistically significant difference in the cost premium by project type; while the alternative hypothesis H_1 is that there is a statistically significant difference in the cost premium by project type. Because the one-way ANOVA test does not show which specific building types significantly differ, the Tukey Post Hoc (TPH) test was subsequently performed to further analyze the difference. Table 8.4 summarizes the results.

TABLE 8.5 ANOVA and TPH Results By Project Size

Project Type	*P*-Value (ANOVA)	*P*-Value (Tukey Post Hoc)	
Commercial projects	.010	Small vs medium	.888
		Medium vs large	.008
		Small vs large	.124
Office projects	.010	Small vs medium	.059
		Medium vs large	.009
		Small vs large	.428
Residential projects	.000	Small vs medium	.875
		Medium vs large	.000
		Small vs large	.000

The P-values from the ANOVA test for projects costing less than S$50 million were smaller than .05, indicating that the hypothesis H_0 should be rejected at 95% confidence level. This result implies that the building type has a significant effect on the mean of the green cost premium when the project size was small or medium. When the project size was large, the building type did not have a statistically significant effect on the mean of the green cost premium.

According to the results from the TPH test, the means of the cost premiums were statistically different between commercial and residential building projects, and between office and residential building projects when the project size was small. As for medium size projects, only commercial and residential building projects had statistically different cost premiums.

From the perspective of project size, large scale projects have the lowest means of green cost premiums in all three building types, followed by small and medium scale projects, as shown in Table 8.5. One possible explanation for this result is that respondents involved in large scale projects were mainly professionals with sufficient experience in green construction projects. Professionals who have sufficient green building experience are able to efficiently utilize green products without increasing the overall design and construction costs (Malin, 2000). Additionally, the cost of green buildings might not increase if the right strategies were used (Bordass, 2000). Respondents with sufficient experience in green building were more likely adopting the right strategies, lowering the green cost premiums accordingly. Another reason derived from the postsurvey interviews was the economies of scale for green features, products, and material.

A one-way ANOVA test was performed to test whether the project size has a significant effect on the green cost premiums. The null hypothesis H_0 is that there is no statistically significant difference in the cost premiums by project size; while the alternative hypothesis H_1 is that there is statistically significant difference in the cost premiums by project size.

The results generated from the ANOVA and TPH tests are shown in Table 8.5. The *P*-values from the ANOVA test for three building types were all smaller than .05, indicating that the hypothesis H_0 should be rejected at 95% confidence level. This result implied that the project size had a significant effect on the mean of green cost premiums regardless of the building type.

Further analysis based on the *P*-value from the TPH test indicated that the green cost premiums were statistically different for medium and large scale projects in all three building types, while the cost premiums were statistically different for small and large scale projects only in residential projects.

The reasons for the cost premium of green construction projects were also investigated. The one sample *t*-test was performed to determine whether each of the reasons had a significant effect on the difference in the cost premiums between green and conventional construction projects. Because a five-point scale was used, the test value was set at 3, which is the middle value of the scale. Table 8.6 summarizes the test result as well as the ranking of the reasons.

With the analysis result, it can be concluded that R1, R5, and R6 had statistically different means from the test value of 3 because the *P*-values of these reasons were all $<.05$. As the mean of R5 (3.70) was much greater than 3, R5 had a significant effect on the difference, however, because the means of

TABLE 8.6 Ranking of the Reasons for the Difference in Cost Premiums

SN	Reasons	*P*-Value	Mean	Rank
R1	Higher consultant and designer fees	.022	2.60	6
R2	Lack of required green expertise and information	.315	3.10	3
R3	Difficulty in getting green services from contractors and subcontractors	.5000	3.00	5
R4	Difficulty in getting green resources e.g., materials, technologies etc.	.444	3.03	4
R5	High cost of green technologies and materials	.000	3.70	1
R6	Lack of Government incentives/ subsidies for green building projects	.034	2.57	7
R7	Higher research and development costs for green building products, systems, technologies etc.	.221	3.13	2

R1 and R6 were much lower than 3, the effects of R1 and R6 could not be considered significant. As the P-values of R2, R3, R4, R5 and R7 were >.05, their mean values were statistically considered at the test value of 3, implying that these reasons could be relevant reasons that could explain the difference as well.

These analysis results suggest that "high cost of green technologies and materials" is the top reason for the difference in the cost premiums between green and conventional construction projects. As the design and construction practices of green buildings became more complex, green technologies and materials not only greatly affected the capital cost, but also the project productivity (Lam et al., 2010). Moreover, green materials usually have higher production costs as these materials lack economies of scale as well as requiring special orders and manufacturing (Malin, 2000; Kibert, 2008). All the interviewees who participated in the postsurvey interview also agreed with this result.

"High research and development costs for green building products and systems" was ranked second, further contributing to the difference in the cost premium. New green products and systems usually require more effort in testing and code approvals, which leads to an increase in the research and development costs (Malin, 2000). "Lack of required green expertise and information," which was ranked third, could also lead to an unnecessary increase in the cost premiums as the key building players are unable to utilize green products efficiently (Malin, 2000). Additionally, without sufficient green building expertise, key building players most probably encounter reworks and changes because they have difficulty in complying with the green standards, leading to an increase in the capital cost for green building projects (Architecture Week, 2001).

"Lack of government incentives/subsidies for green building projects" is not a significant reason. This finding coincides with the results from Hwang & Tan (2012), in which the lack of government support (e.g., incentives) is not an obstacle encountered in green building projects. This is most probably due to the extensive efforts made by the government to support the built industry in Singapore. "Higher consultant and designer fees" was not rated as a significant reason, perhaps because it is not the root cause or a direct reason. Another possible reason may be that getting specialized consultants and designers is not difficult and the cost is not very high (Architecture Week, 2001).

8.3.2 Cost Performance: Green vs Conventional Construction Projects

Table 8.7 summarizes the cost performance of conventional and green construction projects by project type. The negative and positive percentages indicate an "Under Budget" and "Over Budget" cost performance of projects, respectively.

Using the mid-values of the four ranges of the performance, −7.5%, −2.5%, 2.5%, and 7.5%, the mean cost growth of conventional and green construction projects by project type were calculated and are depicted in Fig. 8.2. As can be seen from Fig. 8.2, the conventional projects had negative mean cost growth.

TABLE 8.7 Cost Performances of Conventional and Green Construction Projects

Cost Growth (CG)	No. of Commercial Projects		No. of Offices Projects		No. of Residential Projects	
	Traditional	*Green*	*Traditional*	*Green*	*Traditional*	*Green*
$-10\% \leq CG < -5\%$	10	0	7	0	30	0
$-5\% \leq CG < 0\%$	60	6	4	0	42	0
$0\% \leq CG < 5\%$	54	22	7	1	21	32
$5\% \leq CG < 10\%$	0	29	2	9	5	22
Total	124	57	20	10	98	54

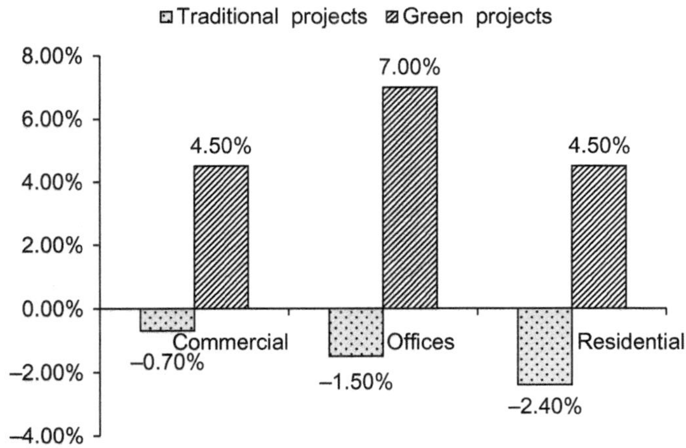

FIG. 8.2 Comparison of the mean cost growth for traditional and green projects. *(Green Building Construction Projects in Singapore: Cost Premiums and Cost Performance, Project Management Journal, 2017. Copyright and all rights reserved. Material from this publication has been reproduced with the permission of PMI.)*

In contrast, the green projects had positive mean cost growth for all three types of projects. This result means that green projects generally had a cost overrun whereas conventional projects were generally under budget. The cost overrun of green construction projects first could be due to the key building players' unfamiliarity and insufficient expertise in green construction projects compared to that in conventional projects. Another possible reason for the cost overrun of green construction projects is that green projects were more likely to be delayed than conventional projects. Hwang and Leong (2013) found that 33.33% of green projects encountered a delay whereas only 17.39% conventional projects did. Furthermore, green office projects had the highest mean cost overrun among the three types of projects. One possible reason is that the respondents were unfamiliar with green office projects, which may be inferred from the small data set for green office projects.

In order to statistically identify the equality in cost growth between green and conventional projects, the independent sample t-test for mean and Levene's Test for variance were carried out. The results are summarized in Table 8.8.

TABLE 8.8 Test Results for Cost Growth

	P-Value	
Project Types	*Levene's Test (Variance)*	T-*Test (Mean)*
Commercial	.315	.000
Office	.000	.000
Residential	.071	.000

The means of cost growth between green and conventional projects in all three types of projects were statistically different because all *P*-values under the *t*-test were $<.05$. Moreover, the variances of cost growth between green and conventional projects in office and residential types of projects were statistically different. Overall, it may be concluded that the cost performance between green and conventional construction projects were statistically different. One reason for the difference could be that the management of green construction projects was challenging due to the complexity of the green technologies and green requirements.

The findings of this survey have alerted practitioners in the green construction industry to pay further attention to cost management. They should adopt proper strategies, tools, and techniques to manage the cost performance of green construction projects from the perspective of time and cost.

8.3.3 Solutions for Cost Premiums Reduction and Cost Performance Improvement

Table 8.9 shows the assessment results of the solutions to reduce cost premiums and improve cost performance. The one sample *t*-test was performed to determine the effectiveness of the solutions to reduce cost premiums and improve cost performance of green building projects. Because a five-point scale was used, the test value was set to 3.

According to the test results, the mean values for the solutions S1 (3.7), S5 (3.83), S6 (3.80) and S8 (3.47) were statistically greater than the test value of 3 because all *P*-values were $<.05$. Moreover, all solutions except S2 and S3 may be considered relevant for reducing cost premiums and improving the cost performance of green building projects because their mean values were statistically equal to or greater than the test value.

"Tax relief," which was ranked as the most powerful solution, is a flexible and feasible solution. It can be given to both businesses and individuals who make the effort to use green products and systems (Bourgeois et al., 2010). This was confirmed by the postsurvey interview. The interviewees agreed that tax relief was a more specific solution that can directly benefit shareholders.

"Availability of skilled and experienced project team" was ranked second, which is consistent with Jiang (2010). Green construction projects generally have a more complex design as compared to conventional construction projects (Johnson, 2000). With a skilled and experienced project team, both lower cost premiums and better cost performance can actually be achieved because the right green design features and materials can be correctly and efficiently adopted during the design and construction period (Hydes and Creech, 2000; Malin, 2000). Furthermore, if a project team has sufficient green building expertise, the cost performance of green construction projects can be significantly improved because unnecessary costs caused by rework and changes could be avoided (Architecture Week, 2001).

TABLE 8.9 Ranking of the Solutions to Reduce Cost Premiums of Green Construction Projects

SN	Solutions to Reduce Cost Premiums	*P*-Value	Mean	Rank
S1	Government to provide incentives/ subsidies for green building projects	.000	3.70	3
S2	Low interest loans	.242	2.83	7
S3	Financial institutions to introduce lending schemes customized for green building projects	.173	2.77	8
S4	Government to provide subsidies for research and development of green building products, systems and technologies	.109	3.33	5
S5	Tax relief for developers and contractors for use of green building products, systems and technologies	.000	3.83	1
S6	Availability of skilled and experienced project team and contractors	.000	3.80	2
S7	Government to provide green building educational courses for key building players so as to flatten the learning curve of green construction	.116	3.30	6
S8	Government to provide subsidies for green building professional and specialist courses	.038	3.47	4

"Incentives/subsidies for green building projects" and "Subsidies for green building professional and specialist courses from the government" were ranked third and fourth, respectively. Practically, incentives from the government are extremely important to attract and motivate hesitant building professionals to build green (Popovec, 2006). Such incentives could enlarge the green market. Additionally, good education in green products and systems can increase the rate of learning in the use of these products. Both cost premium reduction and cost performance improvement may be achieved if the building players are familiar with these green products and systems (Nalewaik and Venters, 2010). The results also coincide with the view from Ong (2013) who pointed out that these green courses and educational programs can be government funded.

8.4 SUMMARY

Green buildings are becoming increasingly popular worldwide. However, in spite of the benefits of green buildings and various efforts made to promote a sustainable built environment, the delivery of green buildings is still hindered by the higher cost associated with "going green." The current cost premiums of green construction projects, and the identification of the significant reasons for these premiums, were investigated in this chapter. In addition, the cost performance of green and conventional construction projects were compared and some feasible solutions to reduce the cost premiums and improve the cost performance were proposed.

The first finding of this survey was that the majority of the respondents perceived the green cost premiums to be 5%–10% with green residential buildings having the highest cost premiums, followed by green commercial and green office buildings. Furthermore, it was proven that "project type" and "project size" were statistically significant variables that affected the cost premiums. It also identified that "high cost of green technologies and materials," "higher research and development costs for green building products, systems, technologies etc.," and "lack of required green expertise and information" were the top three reasons for the cost premiums of green construction projects. As for the current cost performance, it was concluded that green construction projects were generally over budget (4.5%–7%), which was worse than for conventional building projects. Lastly, "tax relief" was identified as the most efficient solution that could have a significant impact on reducing cost premiums and improving the cost performance of green building projects.

This chapter has provided an in-depth understanding of the cost premium and control solutions in green construction projects for both practitioners and researchers. Key building professionals can make better cost-related decisions right at the beginning of green construction projects based on the finding in this chapter. In addition, although the findings appear to be geographically specific to green buildings in Singapore, they should not be limited to the context of Singapore, as Singapore has been globally recognized as one of the leading countries advocating sustainability of the built environment through green buildings. From this perspective, the findings in this chapter will have important implications for the existing body of knowledge as well as for the global construction industry.

Chapter 9

Schedule Performance and Improvement of Green Construction Projects

9.1 INTRODUCTION

Strengthened by the booming economy, construction demand in Singapore remains sturdy. New houses are appearing across the island to satisfy pent-up demand and new offices are being constructed as Singapore enhances its prestige as a regional center for trade and commerce. New facilities are being rolled out to boost the quality of life in a highly dense environment. According to the Building and Construction Authority (BCA, 2014c), Singapore's construction demand reached a historic high of S$35.8 billion in 2013 and will remain strong for the next few years.

With mounting construction demand in a small country like Singapore, sustainable development is not an option but a necessity. Being a city state inhabited by over 5 million residents, and with a land area of about 700 km^2, the challenge lies in attaining the right balance between economic growth and environmental sustainability. Thus, Singapore is moving toward a sustainable development that places a lot of emphasis on green building projects. Singapore has launched a series of green initiatives, which were discussed in Chapter 2, and aims to have 80% of all buildings Green Mark certified by 2030 (BCA, 2009).

In order to achieve this goal, it is important to reduce delays in green construction projects and to ensure that the projects are completed on time. Also, assurance of project schedule is an important indicator of project success, and factors associated with project schedule have been recognized as critical to project success (Ling et al., 2009; Hwang et al., 2013c). If a project can be completed on time or even ahead of time, the overhead costs, labor costs, and interest costs will be reduced. Simultaneously, early completion enables clients to make a higher profit as revenue can be earned earlier.

Construction delays have been widely investigated in previous studies relating to construction management (such as Long et al., 2008; Sweis et al., 2008; Assaf and Al-Hejji, 2006). However, few studies have attempted to analyze the schedule performance of green construction projects. The specific objectives of this chapter are to: (1) investigate schedule performance of green construction

Performance and Improvement of Green Construction Projects. https://doi.org/10.1016/B978-0-12-815483-0.00009-0

projects; (2) identify the critical factors that influence the schedule performance of green construction projects; and (3) provide solutions to improve schedule performance of green construction projects.

9.2 METHOD AND DATA COLLECTION

In this chapter, a questionnaire survey is adopted to collect data, as elaborated in Chapter 3. Prior to developing the questionniare, presurvey interviews were conducted with three project managers from consultants and contractors to validate the schedule performance factors identified from the literature review, as listed in Table 9.1.

The finalized questionnaire consisted of four sections capturing: (1) the profile of the participating companies and respondents; (2) information on conventional and green construction projects performed by companies; (3) the assessment of schedule performance of green construction projects; (4) the critical schedule performance factors; and (5) the solutions for improving the schedule performance of green construction projects.

To investigate the schedule performance, the respondents were asked to provide the total number of conventional projects they had performed in the last 5 years, followed by the number of projects that had experienced schedule delays. For the green construction projects, the respondents provided the same. Also, the respondents selected an appropriate rating based on a defined scale to reflect the importance level of each of the listed factors. The scale was based on a five-point Likert scale with "1" being "not important" and "5" being "very important." Similarly, the five-point Likert scale was also used to evaluate the effectiveness of the recommended solutions to improve the schedule performance of green construction projects.

The questionnaires were sent out to the BCA registered contractors and licensed builders, BCA green mark architectural firms, and Singapore Institute of Surveyors and Valuers. As a result of the survey, 30 completed questionnaires were returned and analyzed (Response Rate = 30%). Thirty respondents who represented their companies participated in the survey, based on the projects that their companies performed during the given timeframe. Due to confidentiality restrictions with the participating companies, the direct collection of individual project performance data was restricted, thus the respondents reported only aggregated data at the company level. Tables 9.2 and 9.3 summarize the profiles of the companies, respondents, and projects.

In addition to the questionnaire survey, interviews with industry experts were conducted to further supplement and substantiate the survey results. Six project managers, who have more than 3 years of experience in managing green construction projects that had been awarded Green Mark Gold Plus and Platinum, were engaged. The analysis results presented in the following section were reviewed by the interviewees. Their shared views and experiences provided deep insight and understanding into the similarities and differences between

TABLE 9.1 Schedule Performance Factors From Literature Review

List of Factors		Reference						
		Male (2003)	*Mansfield et al. (1994)*	*Chan and Kumaraswamy (1996)*	*Majid McCaffer (1998)*	*Assaf and Al-Hejji (2006)*	*Alaghbari et al. (2007)*	*El-Razek et al. (2008)*
Project-related factors	Project delivery methods	√				√		
	Project cost		√	√				√
	Speed of decision making involving all project teams	√		√				
	Communication/ coordination between key parties	√	√	√				√
	Disputes/conflicts between key parties			√		√		
Client-related factors	Speed of decision making by client					√		√
	Delay in progress payment by client				√	√		√
	Client's level of experience in the construction industry	√			√			
	Client initiated variation of works	√		√		√		√

Continued

TABLE 9.1 Schedule Performance Factors From Literature Review—cont'd

List of Factors		Reference						
		Male (2003)	*Mansfield et al. (1994)*	*Chan and Kumaraswamy (1996)*	*Majid McCaffer (1998)*	*Assaf and Al-Hejji (2006)*	*Alaghbari et al. (2007)*	*El-Razek et al. (2008)*
Design team-related factors	Mistakes and delay in producing design documents	√	√	√		√		√
	Level of design team experience	√		√		√		
	Complexity in project design			√		√		
	Misunderstanding of client's requirements by design team	√				√		
Consultant-related factors	Delay in performing inspection and testing		√	√		√		√
	Delay in approving major changes in the scope of works		√			√		√
	Time for reviewing and approval of design documents by consultants			√		√		

	Conflict between consultants and design engineers					√		
	Level of experience of consultants					√	√	
Contractor-related factors	Poor site management and supervision	√		√		√		
	Contractor's deficiencies in planning and updating schedule plans			√		√		√
	Difficulties in financing project by contractors		√			√	√	
	Construction methods implemented by contractors		√			√		
	Rework due to defects during construction		√			√	√	√
Labor-related factors	Shortage of labor			√		√		
	Low labor productivity			√		√		√
	Unskilled labor			√	√		√	

Continued

TABLE 9.1 Schedule Performance Factors From Literature Review—cont'd

List of Factors		Reference						
		Male (2003)	*Mansfield et al. (1994)*	*Chan and Kumaraswamy (1996)*	*Majid McCaffer (1998)*	*Assaf and Al-Hejji (2006)*	*Alaghbari et al. (2007)*	*El-Razek et al. (2008)*
Equipment/ material-related factors	Equipment breakdown			√	√	√		
	Unskilled operators					√		√
	Low productivity and efficiency of equipment			√		√		√
	Lack of high technology mechanical equipment	√				√		
	Availability of material		√	√		√	√	√
	Changes in materials during construction			√		√		√
	Imported materials		√					

External factors	Unforeseen ground conditions	√	√	√		√		√
	Unfavorable weather condition on construction activities		√			√		√
	Accidents during construction					√		√
	Changes in government regulations and law					√	√	
	Delay in performing final. Inspection and certification by third party					√		
	Delay in obtaining permits from political units/ body of officials	√				√		√

TABLE 9.2 Profile of Companies and Respondents

		Years of Experience								
		1–2 Years		*2–3 Years*		*3–4 Years*		*>4 Years*		
	Characteristics	*C-I*	*G*	*C-I*	*G*	*C-I*	*G*	*C-I*	*G*	**Total (*N* = 30)**
Company	Construction	0	2	0	3	0	4	18	9	18
	Consulting	0	0	0	0	0	1	5	4	5
	Development	0	0	0	0	0	1	3	2	3
	Quantity surveying	0	0	0	0	0	0	4	4	4
	Total	0	2	0	3	0	6	30	20	30
Respondent	Project manager/construction manager	0	3	0	3	0	5	18	7	18
	Project director/higher Management	0	0	0	1	0	3	4	0	4
	Project personnel	0	0	0	1	1	2	7	5	8
	Total	0	3	0	5	1	10	29	12	30

Note: "C-I" refers to years of experience in construction industry; "G" refers to years of experience in green construction industry; higher management includes executive president, project director, general manager and managing director; project personnel includes project engineers, construction engineers, and quantity surveyors.

TABLE 9.3 Profile of Projects

Characteristics		Traditional Projects		Green Projects	
		No.	%	*No.*	%
Project type	Commercial	48	21.81	28	29.17
	Residential	91	41.36	21	21.88
	Educational	81	36.81	47	48.96
Project nature	New construction	184	83.63	93	96.88
	Addition and alteration	36	16.36	3	3.12
Project size	Less than S$5 M	32	14.54	3	3.13
	S$5 M to less than S$10 M	6	2.72	8	8.33
	S$10 M to less than S$20 M	35	15.90	6	6.25
	S$20 M to less than S$30 M	33	15.00	13	13.54
	S$30 M to less than S$40 M	15	6.81	16	16.67
	S$40 M to less than S$50 M	27	12.27	18	18.75
	S$50 M and above	72	32.72	32	33.33
Total		220	100.00	96	100.00

green construction and conventional construction. The interviewees also helped in proposing recommendations to improve the schedule performance of green construction projects.

Using the data obtained from the survey, analyses on frequency, descriptive mean scores, and Spearman's ranked correlations, were performed to determine the degree of schedule delays in conventional and green construction projects, as well as the relative association of the ranking of different schedule performance factors. In addition, the Spearman rank correlations were calculated and tested to see if the relative importance of the schedule performance factors of green building projects is significantly correlated with that of conventional construction projects. The test was performed at a confidence level of 0.05.

9.3 ANALYSIS RESULTS AND DISCUSSIONS

9.3.1 Schedule Performance of Green Construction Projects

The companies were asked to input the number of conventional and green construction projects that they had performed as well as the number of projects delayed. The analysis further broke down the projects based on project type, nature, and size to understand how the characteristics of projects affect the project schedules.

As seen in Table 9.4, the overall percentage of projects that had been delayed were 15.91% and 32.29% for conventional and green construction projects, respectively. It can be inferred that green construction projects have a higher probability of delay as compared to conventional construction projects. As green construction projects require green technologies, which are still relatively new to the industry, they require more time to understand and to incorporate these technologies into the design. This is further supported by Snell and Callahan (2005) who pointed out that green building construction takes much longer to complete and presents more challenges than anticipated even if builders are highly experienced at construction, design, and creative problem-solving.

In terms of specific project type, commercial building projects accounted for the highest frequency of delay for both the traditional (25.00%) and green (57.14%) building projects. This may be due to the fact that commercial projects are relatively unique in nature and vary widely in terms of requirements, design, and specifications when compared to residential or educational buildings that can be easily modeled based on past projects. Comparing green and conventional construction, the percentage of delay for green commercial projects (57.14%) was more than twice that of conventional commercial projects (25.00%). One possible reason for a higher delay occurrence in green commercial construction projects is the complexity of incorporating green technologies into the building systems.

Table 9.4 also presents the percentage of projects that were delayed according to project nature, namely, new construction and addition/alteration. It can be seen that more new construction projects were delayed than addition and alteration projects for both conventional and green construction projects. This result could be expected as new construction projects have higher levels of uncertainty than addition and alteration projects because of the wider scope of the projects. It is also observed that new green building construction (33.33%) may encounter a higher likelihood of delay than conventional building construction (17.39%). As green construction is still relatively new to the construction industry, it may be more difficult for players in the industry to manage these types of green construction projects.

Furthermore, large-sized projects, in terms of contract amount, had a higher tendency to encounter project delay. Those projects costing $50 million and

TABLE 9.4 Project Delay: Conventional vs Green Construction Projects

Characteristics		No. of Conventional Projects Performed	No. of Projects Delayed	Percentage Delayed (%)	No. of Green Projects Performed	No. of Projects Delayed Among	Percentage Delayed (%)
Project type	Commercial	48	12	25.00	28	16	57.14
	Residential	91	9	9.89	21	0	0.00
	Educational	81	14	17.28	47	15	31.91
Project nature	New construction	184	32	17.39	93	31	33.33
	Addition and alteration	36	3	8.33	3	0	0.00
Project size	Less than $5M	32	4	12.50	3	1	33.33
	$5M to less than $10M	6	0	0.00	8	2	25.00
	$10M to less than $20M	35	5	14.29	6	0	0.00
	$20M to less than $30M	33	0	0.00	13	3	23.08

Continued

TABLE 9.4 Project Delay: Conventional vs Green Construction Projects—cont'd

Characteristics	No. of Conventional Projects Performed	No. of Projects Delayed	Percentage Delayed (%)	No. of Green Projects Performed	No. of Projects Delayed Among	Percentage Delayed (%)
$30M to less than $40M	15	1	6.67	16	5	31.25
$40M to less than $50M	27	4	14.81	18	7	38.89
$50M and above	72	21	29.17	32	13	40.63
All projects	220	35	15.91	96	31	32.29

above reported the highest percentage of delay for both conventional (29.17%) and green construction projects (40.63%). With more project parties involved in large construction projects, relationships and the flow of information among project team members become more complicated, and inadequate and ineffective coordination can increase the likelihood of communication breakdown and variation, leading to project delay (Long et al., 2004). Large green construction projects, in particular, may require more effective coordination and collaboration than conventional projects due to the complexity inherent in the use of new processes and technologies, and this aspect is reflected in the analysis result.

In addition, the respondents were asked to provide the percentage of additional time required by green construction projects. All the respondents agreed that green construction projects took more time to complete when compared to similar types of conventional projects, and the average of the additional time taken was about 8% of the entire duration of the conventional building projects considered. This is in line with the findings of Glavinich (2008), which stated that green construction projects would take a longer time to complete.

Furthermore, as shown in Table 9.5, the survey results revealed that design development of green construction projects took 17.15% more time than that of conventional projects. Planning and feasibility studies also required an additional time of 15.7%. According to Kubba (2010), green buildings are progressively incorporating more advanced and intricate systems of interacting elements. Furthermore, as mentioned by the project managers interviewed, during the design development stage, greater effort is made to ensure that all the green features are incorporated into design items, and confirmation of the design items is critical. Also, design consultants may require more time to convince the client to adopt a particular green system.

For planning and feasibility studies of green projects, it is necessary to better understand the green technologies to be used, and these may be relatively new to the project team members. Some interviewees reported that most of the green technologies are imported from overseas and to ensure compatibility and performance in the local context, numerous tests must be conducted. As a result, more time must be allocated to the planning and feasibility stage. Furthermore, "Selection of project team" was ranked third as shown in Table 9.5, which indicates that this phase required 14.20% more time than that of conventional projects. This might be because project members involved in green projects need to have appropriate knowledge and experience in green technologies. To select the right team members and to organize an optimized project team, rigorous selection procedures, with a set of qualification criteria, should be developed and implemented, even if more time is required.

The analysis results discussed above were also supplemented by analyzing the impact of the seven phases, which are listed in Table 9.5, on the entire schedule performance of green construction projects. Assuming that those phases requiring more time may have a stronger impact on the project schedule performance, the survey asked the respondents to rank the level of relative

TABLE 9.5 Green Project Duration and Schedule Impact

Project Phase	Additional Duration of Green Project (% of Traditional Project Duration)		Impact on Green Construction Schedule Performance	
	Mean	*S.D.*	*Mean Rank*	*S.D.*
Planning and feasibility studies	15.70	7.07	3.43	2.25
Selection of project team	14.20	8.18	4.30	2.31
Schematic design	9.62	7.22	3.10	1.13
Design development	17.15	7.15	2.23	1.31
Tender document	9.37	6.72	4.70	1.09
Construction	12.20	7.93	4.30	1.56
Closing out and commissioning	3.77	2.58	5.93	1.74

significance of the phases, with 1 being most significant and 7 being least significant. The analysis revealed that the design development stage had the highest impact on schedule performance of green projects with mean rank of 2.23. In addition, the schematic design (mean rank = 3.10) and the planning and feasibility study (mean rank = 3.43) were ranked second and third, respectively. Based on these results, it may be deduced that project planning, feasibility studies, design development, and schematic design need to be adequately managed to produce good schedule performance for green construction projects.

In order to identify the schedule performance of green construction projects, the frequency and the percentage of schedule overrun were analyzed, and Table 9.6 presents the summary of the analysis results. The respondents were first required to report the frequency of green construction projects completed later than the as-planned schedule. As shown in Table 9.6, 50% of the respondents (15) indicated that half of the green construction projects that their companies performed were completed behind their planned schedule. On average, the frequency of schedule overrun of green construction projects was approximately 43%, indicating that about 4 out of 10 green projects that they had performed experienced schedule overrun. This probability is quite high and it may be attributable to the complexity of green construction. Another possible reason is that more time is required by key project team members to fully understand the green technology systems that are incorporated into the design. In addition, it is necessary to ensure that the green elements are able to be integrated with the

TABLE 9.6 Schedule Overruns of Green Building Construction Projects

	Frequency of Schedule Overrun of Green Building Construction Projects				Schedule Overrun of Green Building Construction Projects (% of Planned Schedule)				
Scale	*No. of Res.*	%	*Mean (%)*	*S.D.*	*Scale (%)*	*No. of Res.*	%	*Mean (%)*	*S.D.*
Not at all (0%)	3	10.00	43.33	21.71	0	3	10.00	4.80	2.76
Seldom (1%–49%)	7	23.33			1–3	5	16.67		
Half of the time (50%)	15	50.00			4–6	16	53.33		
Most of the time (51%–99%)	5	16.67			7–9	3	10.00		
All the time (100%)	0	0.00			≥10	3	10.00		

other building systems fully to prevent any conflicts during construction (Glavinich, 2008; GreenBiz, 2005; Kats et al., 2003).

Furthermore, as shown in Table 9.6, more than half of the respondents (16) reported that the magnitude of schedule overrun of their projects ranged from 4% to 6% of the as-planned schedule. Also, the mean shown in Table 9.6 (4.8%) indicated that the actual schedule of the green projects was delayed by an average of 4.8%, which might be considered relatively severe. Thus, it can be of paramount concern to both clients and contractors as such delay has a substantial effect on the overall project performance, ultimately reducing the chance of producing successful projects.

9.3.2 Critical Factors Affecting Schedule Performance of Green Construction Projects

Having investigated the degree of project delay incurred in conventional and green construction projects, it became apparent that identifying various factors responsible for project delay is necessary to improve schedule performance. As a result, the relative importance of the factors regarded by the respondents as major causes of project delay was analyzed for conventional and green construction projects. The analysis results are presented in Table 9.7.

9.3.2.1 Project-Related Factors

In conventional construction projects, factor P4 yields the highest mean score of 4.53 as seen from Table 9.7. It can be interpreted that the respondents considered communication and coordination between key parties as the most important factor that significantly affects the project schedule. Because factor P3, with mean score of 4.53, is relatively close to the mean score of factor P4, it can also be derived that the speed of decision making, involving all project teams, is also a critical factor that may affect the project schedule. Factors P1 and P2 are ranked an average of close to a mean of 4.0 showing a degree of importance as well. However, factor P5 has the lowest rank at a mean score 3.17, which signified that legal disputes and conflicts between key parties were of minor importance in the project schedule.

For a green construction project, the most important factor in causing delays to the project was factor P3 with a high means score of 4.77. Although factor P4 is second in rank with a mean score of 4.70, the close mean scores between the two factors indicate that factor P4 is as important as factor P3. Similar to conventional projects, P1 and P2 are ranked with an average of 4.30, while factor P5 is lowest in rank denoting that it has the lowest impact on the project schedule.

The third, fourth, and fifth rank of project-related factors are consistent between conventional and green construction projects. In contrast, the first and second ranks of the project-related factors for the two groups are inverted. Because the top two factors between conventional and green construction

TABLE 9.7 Schedule Performance Factors: Conventional vs Green Construction Projects

Factors			Within Categories						Cross Categories			
			C		*G*				*C*	*G*		
			M	*R*	*M*	*R*	*S*	P-*Value*	*OR*	*OR*	*S*	P-*Value*
Project-related factors	P1	Project delivery methods	4.03	3	4.30	3	0.895	.040	18	16	0.913	.000
	P2	Project cost	4.03	4	4.30	4			18	16		
	P3	Speed of decision making involving all project teams	4.53	2	4.77	1			3	2		
	P4	Communication/coordination between key parties	4.57	1	4.70	2			2	3		
	P5	Disputes/conflicts between key parties	3.17	5	3.47	5			39	35		
Client-related factors	CL1	Speed of decision making by client	4.53	1	5.00	1	1.000	.000	3	1		
	CL2	Delay in progress payment by client	4.07	3	4.17	3			17	20		
	CL3	Client's level of experience in the construction industry	4.43	2	4.67	2			7	6		
	CL4	Client initiated variation of works	3.80	4	4.13	4			26	23		

Continued

TABLE 9.7 Schedule Performance Factors: Conventional vs Green Construction Projects—cont'd

Factors			Within Categories						Cross Categories			
			C		*G*				*C*	*G*		
			M	*R*	*M*	*R*	*S*	P-*Value*	*OR*	*OR*	*S*	P-*Value*
Design team-related factors	DT1	Mistakes and delay in producing design documents	4.47	1	4.27	3	0.400	.600	5	18		
	DT2	Level of design team experience	4.03	3	4.63	1			18	7		
	DT3	Complexity in project design	4.13	2	4.53	2			14	10		
	DT4	Misunderstanding of client's requirements by design team	4.00	4	4.17	4			22	20		
Consultant-related factors	CS1	Delay in performing inspection and testing	4.13	5	4.00	5	0.700	.188	14	27		
	CS2	Delay in approving major changes in the scope of works	4.47	2	4.57	3			5	9		
	CS3	Time for reviewing and approval of design documents by consultants	4.60	1	4.63	2			1	7		

	CS4	Conflict between consultants and design engineers	4.17	4	4.43	4			13	14		
	CS5	Level of experience of consultants	4.43	3	4.70	1			7	3		
Contractor-related factors	CO1	Poor site management and supervision	4.20	3	4.53	2	0.975	.005	11	10		
	CO2	Contractor's deficiencies in planning and updating schedule plans	4.20	2	4.47	3			11	12		
	CO3	Difficulties in financing project by contractors	4.40	1	4.70	1			9	3		
	CO4	Construction methods implemented by contractors	3.93	4	4.27	4			24	18		
	CO5	Rework due to defects during construction	3.27	5	3.43	5			37	36		
Labor-related factors	L1	Shortage of labor	4.13	1	4.10	1	1.000	.000	14	25		
	L2	Low labor productivity	3.63	2	3.67	2			30	30		
	L3	Unskilled labor	3.50	3	3.57	3			34	31		

Continued

TABLE 9.7 Schedule Performance Factors: Conventional vs Green Construction Projects—cont'd

			Within Categories						Cross Categories			
			C		*G*				*C*	*G*		
Factors			*M*	*R*	*M*	*R*	*S*	P-*Value*	*OR*	*OR*	*S*	P-*Value*
Equipment/ material-related factors	EM1	Equipment breakdown	3.40	6	3.37	7	0.685	.090	36	37		
	EM2	Unskilled operators	3.27	7	3.53	4			37	32		
	EM3	Low productivity and efficiency of equipment	3.57	5	3.50	5			33	34		
	EM4	Lack of high technology mechanical equipment	3.60	4	3.37	6			32	37		
	EM5	Availability of material	3.63	3	4.17	2			30	20		
	EM6	Changes in materials during construction	3.83	2	4.07	3			25	26		
	EM7	Imported materials	4.03	1	4.33	1			18	15		

External factors	E1	Unforeseen ground conditions	3.67	5	3.53	5	0.943	.095	29	32		
	E2	Unfavorable weather condition on construction activities	3.50	6	3.37	6			34	37		
	E3	Accidents during construction	3.97	2	4.13	2			23	23		
	E4	Changes in government regulations and law	3.73	4	3.90	3			28	28		
	E5	Delay in performing final. Inspection and certification by third party	3.80	3	3.87	4			26	29		
	E6	Delay in obtaining permits from political units/ body of officials	4.40	1	4.47	1			9	12		

Note: C, conventional; *G*, green; *M*, mean; *OR*, overall rank; *R*, rank; *S*, Spearman rank correlation coefficient.

projects have mean scores of >4.5, it can be concluded that communication/coordination among key parties is required to ensure that decision making can happen quickly to avoid delays in the project.

To further examine the agreement in the ranking of the project-related factors between conventional and green construction projects, a Spearman's correlation coefficient of .895 was obtained. This shows a positive correlation between the ranking of the project-related factors under conventional and green construction projects because a high value of rank correlation coefficient suggests a strong agreement between the two groups. Because the P-value of .04 at 95% significance level is less than the significance, $\alpha=.05$, this shows that there is a significant relationship between the conventional and green construction projects for factors that are project related. Agreement between the rankings of the factors for both groups is observed as the respondents may feel that the extent of the effect of the individual factors is similar for both conventional and green construction projects.

9.3.2.2 *Client-Related Factors*

With reference to Table 9.7, the ranking of client-related factors between conventional and green construction projects is identical. Respondents agreed that the speed of decision making by the client has the highest effect on schedule delay in this category. Factor CL3, with mean scores of above 4.5 for both conventional and green projects, has a considerable effect on the project schedule as well. Therefore, to ensure positive schedule performance, the client's speed of decision making and level of experience is vital to ensure that the schedule is not delayed.

Under green construction projects, all the client-related factors have mean scores above 4.00. This illustrates the fact that the client's impact on schedule is of great magnitude and should not be overlooked.

The Spearman's rank correlation coefficient of 1.00 indicates a positive correlation between the two groups. Because the P-value of .000 at 95% significance level is less than the significance, $\alpha=.05$, it implies that there is a strong agreement in the ranking of the factors under the client related categories. There is strong agreement on the rankings of the impact of the factors affecting project schedule under the two groups as the level of client involvement might be similar for the two types of projects. It might also be due to the fact that each factor has equal impact on conventional and green project schedules.

9.3.2.3 *Design Team-Related Factors*

Factor DT1 has the highest ranked mean of 4.47 under conventional construction projects. It is followed by factors DT3, DT2, and DT4 respectively. Among all the design team-related factors, mistakes and delays in producing design documents has the highest influence in causing schedule delay. For green construction projects, factor DT2, with mean score of 4.63, is ranked the highest with factors DT3, DT1, and DT4 ranked behind it in that order. "Level of design team experience" is rank highest under green construction projects as team members should be familiar with the specifications of green technology when it is incorporated into the design of the building systems.

The Spearman's rank correlation coefficient of .20 indicates a weak correlation between the two groups. In addition, because the *P*-value of .600 at 95% significance level is more than the significance, $\alpha = .05$, it implies that there is no significant relationship and insubstantial agreement in the ranking of the factors under the client related category between the two groups. As such, we can conclude that the severity of the factors varies between the conventional and green construction projects.

The disparity in the ranking between the two groups may be because more emphasis is placed on the design of green construction projects. According to Kubba (2010), green buildings are progressively incorporating more advanced and intricate systems of interacting elements. During design, the impact of the elements on each system must be considered as a whole. A failure to take into account the integration of green technologies and their impact on other building elements would result in construction conflicts leading to delay in order to address such problems. As such, the level of experience of the design team, as well as the complexity of the design, have the highest impact on the green construction project schedule as compared to a conventional construction project.

9.3.2.4 Consultant-Related Factors

Under conventional construction projects, factor CS3, with a mean of 4.60, is ranked first, followed by factors CS2, CS1, and CS4 in that order. Factor CS5, with a mean of 4.70, is considered to have the highest impact on the green project schedule among all the consultant-related factors. It is similar to that of design team-related factors for green building projects, in which the level of design team experience is considered the factor with the highest impact on project schedule delay.

The Spearman's rank correlation coefficient of .70 indicates a fairly strong correlation between the two groups. However, the *P*-value of .188 at 95% significance level is more than the significance, $\alpha = .05$. This implies that there is insufficient agreement and relationship on the ranking of the factors between the two groups even though the rank correlation coefficient is relatively strong. As such, it is deduced that the severity of the factors varies between conventional and green construction projects.

Level of experience of the consultant team is an important factor under green construction projects as the level of design complexity is higher than that of traditional construction projects. With reference to Kubba (2010), specialist consultants should be involved in the design process earlier in order to incorporate their suggestions and requirements in the design, thus ensuring that their contributions are taken into account to safeguard maximum efficiency. Therefore, without the required level of knowledge and experience by the consultants, harmonization of the systems would not be possible and this would pose an elevated risk of having conflicts. In addition, a delay in reviewing and approving the design would further impede the project schedule.

9.3.2.5 Contractor-Related Factors

With reference to Table 9.7, factor CO3 is deemed as the most influential factor in project schedule delay for both conventional and green construction projects. The mean score of factor CO3 under conventional construction projects is found to be 4.40 and to be 4.70 under green construction projects. Factors CO1 and CO2 are ranked second and third interchangeably for both groups as seen in Table 9.7. It should be noted that, aside from lack of finance for the project, which would cause the project schedule to be delayed, good site management and supervision, accompanied by the ability to plan and monitor the project schedule appropriately, is vital to ensure that the optimal project schedule performance is met.

The Spearman's rank correlation coefficient of .90 indicates a strong correlation between the two groups. Because the P-value of .005 at 95% significance level is less than the significance, $\alpha = .05$, there is adequate agreement and a significant relationship in the ranking of the level of impact of the contractor-related factors between conventional and green construction projects. Both green and conventional projects have similar rankings on the impact of contractor-related because it is believed that the degree of influence of the variables associated with the contractor is unaffected by the type of project.

9.3.2.6 Labor-Related Factors

In this category, the statistics show that there is unanimous agreement by the respondents on the level of impact of the three factors on both conventional and green construction projects. Factor L1 is regarded as having the highest impact on project schedule for both conventional and green construction projects, while factors L2 and L3 are ranked second and third.

The Spearman's rank correlation coefficient of 1.00 indicates a perfect linear correlation between the two groups. Because the P-value of .00 at 95% significance level is less than the significance, $\alpha = .05$, it suggests that there is a significant relationship between the ranking of the factors under both the green and conventional construction projects. There is strong agreement on the rankings of the impact by labor-related factors affecting the project schedule as the influence might be comparable in the two types of projects.

9.3.2.7 Equipment and Material-Related Factors

As seen in Table 9.7, both conventional and green construction projects have ranked factor EM7 as the most important factor that may cause a project schedule delay in this category. Imported materials are long lead items that require a large number of weeks or months to be delivered on site. As such, any technical hitch affecting the delivery of imported materials, especially when it concerns the critical activities in a schedule, would adversely affect the project, causing a delay. With reference to Table 9.7, a majority of the factors have a mean score <4.0, which implies that equipment and material-related factors have relatively less impact on project schedule as compared to other categories. This might be due to the fact that equipment and material-related factors can be better controlled and predicted than factors in other categories.

The Spearman's rank correlation coefficient of .714 indicates a strong correlation between the two groups. Because the *P*-value of .090 at 95% significance level is more than the significance, $\alpha = .05$, there is an insignificant relationship and agreement on the ranking of the factors between the two groups.

Disagreement on the rankings of the equipment and material-related factors is observed between conventional and green construction projects and this might be due to the fact that green technologies are usually imported from overseas. Therefore, the availability and delivery of these technologies is of higher importance (second in rank) than conventional construction projects (third in rank). A low level of equipment-operator skill is ranked fourth under green construction projects; while it is sixth in rank under conventional construction projects. It may be attributed to the fact that green construction projects require extensive knowledge and experience to incorporate the green technologies into the construction design and during the construction. As such, higher emphasis is place on the skill of the operator to construct and integrate the green technologies with other construction systems.

9.3.2.8 External Factors

Factor E6 is first in rank with a mean of 4.40 and 4.70 for conventional and green construction projects respectively. The high mean scores of factor E6, which are around 4.5 for both groups, imply that any impediment in obtaining permits from authorities will likely cause the project to experience a delay. The ranking of factors between the two groups are relatively similar expect for the third- and fourth-ranked factors.

The Spearman's rank correlation coefficient of .943 indicates a strong correlation between the two groups. Because the *P*-value of .005 at 95% significance level is less than the significance, $\alpha = .05$, there is ample agreement and significant relationship on the ranking of the factors between the two groups.

It can be observed that there is an agreement in the rankings of impact of the external factors for both green and conventional projects. This might be because external factors are more unpredictable and it is hard to gauge whether there might be any difference on the degree of impact on the project schedule for different types of project. Therefore, it is assumed that the level of effect is comparable for conventional and green construction projects.

9.3.2.9 Rank for All Factors

The top 10 most significant delay factors for conventional construction projects as identified in the survey were, in sequence, CS3, P4, CL1, P3, DT1, CS2, CL4, CS5, E6, and CO3. For green construction projects, the 10 most significant delay factors were CL1, P3, CO3, CS5, P4, CL4, CS3, DT2, CS2, and DT1.

The Spearman's rank correlation coefficient value of .913 indicates a strong correlation between the two groups. Because the significance value = .000 at 95% significance level is less than the significance, $\alpha = .05$, there is ample agreement and significant relationship in the ranking of the factors between the two groups.

The common top three most important factors that have the greatest effect on project schedule delay for both green and conventional construction projects are (1) speed of decision making by client, (2) speed of decision making involving all project teams, and (3) communication/coordination between key parties.

The speed of decision making by client was ranked first under conventional construction projects, and third under green construction projects. The speed of decision making by client is considered one of the most significant factors as key decisions are ultimately made by the client. Any delay in decision making from the client, especially when the decisions have a strong impact on activities on the critical path, will impinge on the progress and schedule of the project.

Speed of decision making involving all project teams was ranked second for conventional construction projects, and third under green construction projects. It is essential that effective decisions should be taken at the right time. To achieve a successful project, the flow of information between all project team members should be systematic and timely, and should address the appropriate personnel. In addition, key decision makers should be clearly identified to enable effective decision making.

Communication and coordination between key parties was identified as another major factor by the respondents. Under green construction projects, it was ranked second, while it was fifth in rank under conventional construction projects. From Chalabi and Camp (1984), it was found that inadequate communication between project team members was a critical factor causing delays, and that it impeded the progress of the project, leading to problems in project coordination and schedule. Communication and coordination between key parties have a greater effect on schedule delay in green construction projects due to the fact that green construction projects require a more holistic and integrated approach because the design process is more complex and much of the design phases overlap with the construction phases (Glavinich, 2008). As such, green construction projects require more effort in communicating among the various team members (Kibert, 2008).

In Table 9.7, the mean scores for each factor category are computed by taking an average of the mean scores of all the factors under the category. The mean scores for each category are then ranked. The Spearman's rank correlation coefficient of .976 signifies a strong correlation between conventional and green construction projects. Because the *P*-value of .000 at 95% significance level is less than the significance, $\alpha = .05$, there is strong agreement and a significant relationship on the ranking of the factors between the two groups.

As seen from Table 9.7, there is substantial similarity between the top 3 factor categories that are deemed to have the highest impact on project schedule delay for both conventional and green construction projects. The top three factor categories having the greatest influence on project schedule delay include: (1) client related factor, (2) consultant related factor, and (3) design team related factor.

Client, consultant and design team-related factors were regarded to have the greatest impact on project schedule delay as it was agreed by most respondents that the planning and feasibility stage, together with the design stage, were the two most important stages that determined project schedule performance. This was due to the fact that if the two stages were well-planned and prepared, the design documents, such as drawings and specifications, would be more complete and encompass all the requirements of the client. As such, conflicts and changes could be prevented, thus making it possible for the project to complete on time or ahead of time (Al-Momani, 2000). In cases in which the client did not pay progress payments on time, the contractors could not finance the work, and thus clients should minimize variations to prevent (Assaf et al., 1995).

9.3.3 Solutions for Schedule Performance Improvement of Green Construction Projects

The survey inquired about the respondents suggested recommendations to enhance green construction project schedule performance. A total of 14 possible solutions were listed in the survey, as shown in Table 9.8. The top five solutions are also shown. The findings from the survey can be used to identify areas and aspects that should first be addressed in order to improve the schedule performance of green construction projects. Table 9.8 gives a ranking of the solutions based on the frequency of the top five selected recommendations by the respondents.

Based on the result, the top three ranked solutions are discussed as follows:

- *Ensure that the actual construction schedule and resources are seriously monitored and reviewed so that the performance is in line with as-planned to avoid chances of cost overrun and disputes.* Project managers should check that the green construction planning and scheduling are continuous processes during construction and that the schedule plans correspond with the time to develop work and resources to prevent cost overruns and disputes

TABLE 9.8 Solutions for Improving the Schedule Performance of Green Construction Projects

Solutions for Schedule Improvement of Green Buildings	Frequency of Top 5 Selected	Percentage (%)	Rank
Ensure that the actual construction schedule and resources are seriously monitored and reviewed so that the performance is in line with as-planned to avoid the chances of cost overrun and disputes	23	76.67	1
Check for mistakes and discrepancies in design documents to avoid redoing of designs and drawing before submission for approval to avoid variations or necessary corrections	19	63.33	2
Alternative procurement method should be analyzed to ensure it meets the project requirements and complexity	16	53.33	3
Ensure that design documents are produced on time	14	46.67	4
Ensure that there is optimum number of labors to be assigned for individual activities and motivate the workers to increase productivity	14	46.67	5
Pay progress payment to the contractor timely to facilitate the contractors' ability to finance the work	12	40.00	6
To minimize variation orders to avoid delays	11	36.67	7
Contractor should manage his financial resources and plan cash flow by utilizing progress payment	7	23.33	8
Avoid delay in when reviewing and approval of documents, materials or others are needed	6	20.00	9
Ensure that contractor has the capability and the resources for constructing the project before awarding the tender	4	13.33	10
Administrative and technical staff should be appointed as soon as project	4	13.33	11

TABLE 9.8 Solutions for Improving the Schedule Performance of Green Construction Projects—cont'd

Solutions for Schedule Improvement of Green Buildings	Frequency of Top 5 Selected	Percentage (%)	Rank
is awarded to facilitate the project completion within specified time with the required quality and estimated cost			
Consultants should not delay the checking, reviewing and approving of design documents leading to a delay in construction phase	3	10.00	12
Consultant should be flexible in evaluating contractors work and balance between the required quality with cost	3	10.00	13
Contractor should carry out a comprehensive economic analysis and workable financial plans	3	10.00	14

(Assaf et al., 1995). To ensure that the project schedule and resources are constantly monitored and reviewed, owners may deploy certain tools and techniques for schedule control. A schedule control system may be set up to define the procedures when changes occur in the project schedule, and performance measurements should be carried out to assess the degree of variations that took place (Project Management Institute, 2008).

- *Check for mistakes and discrepancies in design documents to avoid redoing of designs and drawing before submission for approval to avoid variations or necessary corrections.* When mistakes and discrepancies are discovered in design documents, redoing of designs and drawing requires extra time beyond the as-planned schedule to make the necessary changes (Assaf et al., 1995) and it will more often than not lead to poor time performance (Chan and Kumaraswamy, 1996). When these mistakes and discrepancies are detected, especially during the construction stage, variations leading to rework may take place, thus giving rise to redundant work with additional cost introduced to make the corrections.
- *Alternative procurement method should be analyzed to ensure it meets the project requirements and complexity.* Every procurement method has a different effect on project development and involves different relationships between all the project team members (Nofera and Korkmaz, 2010). As such,

clients should analyze the project characteristics with care to select the appropriate procurement method as misinterpretation of project characteristics would probably lead to poor project performance cost increases (Council, 2003). Due to the greater complexity of the technical systems used in green construction projects, these projects require high levels of interdependency, communications, and close partnerships with all the project participants during the design stages (Kibert, 2008; Magent et al., 2009). In addition, research has also shown that the early involvement of owners enables green construction projects to achieve their green objectives at lower cost (Beheiry et al., 2006; Lapinski et al., 2006). Accordingly, it is vital to analyze alternative procurement methods to achieve better green construction project performance.

9.4 SUMMARY

Construction schedule performance was closely linked to construction cost and quality, and intrinsically, it was used as a benchmark to evaluate the performance of the entire project and efficiency of the project organization. In addition, green construction projects require a more holistic and integrated approach in their delivery systems as compared to conventional construction projects. As green construction requires a more detailed design due to its complex technological characteristics, greater efforts in communication and collaboration between project parties are needed.

This chapter addressed the schedule performance of green construction projects, identified and assessed the most important factors affecting the schedule performance, and provided some recommendations that may improve the targeted performance. The results first showed that (1) 15.91% of the investigated conventional projects were delayed while 32.29% of the investigated green construction projects were completed behind schedule. In addition, green construction projects required an average of 8% more time than conventional construction projects of similar size and characteristics. On average, these projects were delayed by 4.8% compared to their as-planned schedules. Furthermore, the top five factors causing delay in green projects were: (1) speed of decision making by client; (2) speed of decision-making involving all project teams; (3) communication/coordination between key parties; (4) level of experience of consultants; and (5) difficulties in financing project by contractors. Lastly, a list of solutions for schedule-related issues was introduced. These studies will offer a benchmark for the industry to gauge the overall duration and performance of green construction projects and serve as a basis for further research on the enhancement of green construction schedule performance.

Chapter 10

Safety Performance and Improvement of Green Construction Projects

10.1 INTRODUCTION

Construction work is inherently dangerous because of its unique nature, and construction has been regarded as one of the most hazardous and unsafe industries worldwide, with fatality and injury rates that are significantly higher than those in any other industry (Hinze, 1997; Lingard and Rowlinson, 2005; Patel and Jha, 2015; Guo and Yiu, 2016). Clearly green construction will present a different set of challenges and uncertainties that are unknown in conventional construction, which explicitly undermine the safety integrity in the green construction industry (Fortunato et al., 2011). Hence, examining the issue of safety in green construction projects is crucial.

Current literature and reports show that, compared to conventional construction projects, green construction projects impose a higher safety risk to workers. Fortunato et al. (2011) found that construction projects that were built to achieve LEED certification accounted for a higher recorded injury rate than non-LEED constructions. Walter (2011) reported that workers engaged in LEED projects were exposed to the risk of working at heights for a longer duration because of the numerous construction activities performed on roofs, such as constructing atria and installing solar panels or vegetated roofs. In addition, statistics have shown that workers involved in LEED construction projects experience a 36% increase in lacerations, strains, and sprains from recycling construction materials; a 24% increase in falls to lower levels during roof work because of the installation of on-site renewable energy sources, such as solar panels; a 19% increase in eye strain when installing reflective roof membranes; and a 14% increase in exposure to harmful substances when installing innovative wastewater technologies, such as on-site wastewater treatment systems (Walter, 2011). Therefore, more emphasis should be placed on controlling and eliminating safety concerns in green construction projects than in conventional construction projects.

The early identification of safety issues in construction projects has been acknowledged as vital because it allows stakeholders to comprehend the

Performance and Improvement of Green Construction Projects. https://doi.org/10.1016/B978-0-12-815483-0.00010-7

repercussions of their actions fully, fine-tune their work-related activities, and implement corrective measures to prevent the recurrence of similar accidents. Thus, identifying safety issues is extremely critical to the improvement of safety performance in green construction projects. However, the extant literature shows that little research effort has gone into this field.

The aims of this chapter are (1) to conduct an investigation and comparison of safety performance and critical safety issues between green and conventional construction projects in Singapore, and (2) to propose a series of feasible solutions to improve the safety performance in green construction projects.

10.2 METHOD AND DATA PRESENTATION

In this chapter, a mixed approach of qualitative and quantitative methods, including a literature review, structured interviews, and a questionnaire survey, are used. First, 27 safety issues were identified through a comprehensive literature review and these are listed in Table 10.1. Afterward, structured interviews were conducted to elaborate these safety issues within the context of Singapore. Two BCA certified project managers and one Green Mark manager with at least 25 years of experience in the construction industry and 5 years of experience in the green construction industry were invited to participate in the structured interviews because of their considerable knowledge and experience. Open-ended questions were also asked to obtain constructive responses and feedback.

TABLE 10.1 Safety Issues Identified From Existing Literature

Safety Issue	Source	Frequency
Fall from height	[1], [2], [3], [4], [5], [7], [10], [11], [12], [13], [18], [20], [22], [23], [24], [28]	16
Crush injuries	[1], [27]	2
Fall on same level from slipping, tripping, and stumbling	[1], [4], [5], [30]	4
Being struck by falling objects	[1], [2], [4], [5], [12], [20], [23], [24], [25], [27], [28]	11
Moving/handling heavy loads	[1], [4], [5], [18]	4
Bad working positions (confined spaces)	[1], [19], [20]	3
Being struck/crushed by vehicle	[1], [4], [26]	3

TABLE 10.1 Safety Issues Identified From Existing Literature—cont'd

Safety Issue	Source	Frequency
Electrocution	[1], [2], [3], [4], [5], [6], [7], [8], [9], [12], [29]	11
Trapped/caught in/between objects	[2], [3], [4], [5], [23], [27]	6
Collapse of tower crane	[2], [9], [21], [22]	4
Being struck by moving objects	[2], [3], [4], [5], [9], [23], [24], [28]	8
Being struck by stationary objects	[4], [5], [27], [28]	4
Being struck against manually operated tools	[3], [27], [28]	3
Exposure to heat/cold	[2], [16], [17]	3
Inhalation	[3]	1
Respiratory failure	[3]	1
Fall on same level due to collision with, or pushing by, another person	[3], [4], [28], [30]	3
Contact with/by machinery	[4], [20]	2
Exposure to fire	[5], [12]	2
Exposure to hazardous substances	[5], [14], [15], [27]	4
Contact with heavy vehicles	[12], [20], [26], [29]	4
Traffic accidents on site	[12]	1
Cave-ins	[12]	1
Building/structure collapse	[12]	1
Contact with hand-tools	[20], [23], [27]	3
Foreign matter in eye	[27]	1
Drowning	[29]	1

Note: [1] WSHC (2013), [2] WSHC (2010), [3] Chi and Han (2013), [4] Haslam et al. (2005), [5] Perlman et al. (2014), [6] Still et al. (1997), [7] Chi et al. (2009), [8] Chi et al. (2012), [9] Al-Humaidi and Tan (2009), *[10] (Chi et al., 2005),* [11] Kines (2003), [12] Müngen and Gürcanli (2005), [13] Aneziris et al. (2008), [14] Ilvonen (2013), [15] Lupsea et al. (2014), [16] Rowlinson et al. (2014), [17] Yi and Chan (2013), [18] Azevedo et al. (2014), [19] Spillane et al. (2011), [20] Irumba (2014), [21] Aneziris et al. (2008), [22] Swuste (2013), [23] Tam and Fung (2011), [24] Wu et al. (2013), [25] Aneziris et al. (2014), [26] McCann (2006), [27] Lipscomb et al. (2010), [28] Schoonover et al. (2010), [29] Hinze and Teizer (2011), and [30] World Health Organization (2016).

One safety issue, namely, being struck by stationary objects, was omitted after the respondents reported that this issue barely occurred on site based on their experience. Finally, 26 critical safety issues were consolidated.

A questionnaire was then developed based on the structured interview results. The questionnaire consisted of a cover page with an explanation of the survey objectives, followed by four independent sections. Section A gathers the profiles of the respondents and their companies. Section B counts the number of conventional and green construction projects that have been undertaken by each company. This section also records the safety performance of these projects. Section C evaluates the likelihood of occurrence and the severity of each critical safety issue in conventional and green construction projects using two five-point Likert scales (1 = rare/negligible, 2 = remote/minor, 3 = occasional/moderate, 4 = frequent/major, and 5 = nearly certain/catastrophic). Based on a comprehensive literature search through news, reports, and journal articles (Dewlaney et al., 2011; Fortunato et al., 2011; Walter, 2011; Pearce and Kleiner, 2013; Rajendran et al., 2009; Bortolini et al., 2016; Griffith, 2002; Rechenthin, 2004; Rajendran and Gambatese, 2009; Dewlaney and Hallowell, 2012; Hajmohammad and Vachon, 2014), Section D gathers a collection of solutions that may enhance the safety performance in green construction projects and requests the respondents to endorse those they favor.

The sampling framework of the questionnaire survey was a list of 102 BCA registered construction companies that possess experience in both green and conventional construction projects in Singapore. The number of questionnaires disseminated was determined by two considerations. The first consideration is that statistical analysis can only be conducted when the sample size is >30, because the central limit theorem is inapplicable when the sample size is less than that (Ott and Longnecker, 2015). The second consideration is that the norm response rate for a construction survey is around 30% (Akintoye, 2000). Thus, a total of 100 questionnaires were sent out initially with the expectation of receiving around 30 replies, a sufficient number for statistical analysis. During the data collection, however, two more qualified companies expressed interest in the topic and thus were also included. Hence, in the aggregate, 102 questionnaires were disseminated. In addition, it is worth noting that BCA has a four-level grading system (A1, A2, B1 and B2) for registered construction companies, in which A1 companies enjoy no tendering limit, while A2, B1, and B2 companies can only bid for projects worth up to SGD 85 million (approximately USD 60 million), SGD 40 million (approximately USD 28 million) and SGD 13 million (approximately USD 9 million), respectively (BCA, 2016d). Thus, to represent companies across the spectrum, questionnaires were randomly and evenly disseminated among companies at different levels. The questionnaire was disseminated via email between October and November 2014, and 30 replies were received, yielding a norm response rate of 29.41% for the construction industry survey (Akintoye, 2000). Tables 10.2 and 10.3 profile the companies, respondents, and projects involved in the survey.

TABLE 10.2 Profile of Companies and Respondents

Profile	Number	Percentage (%)
Company (total = 30)		
Type		
Contractor	20	66.67
Consultant firm	4	13.33
Private developer	3	10.00
Quantity surveyor firm	3	10.00
Years of experience in conventional construction projects		
<2 years	0	0.00
2–3 years	0	0.00
3–4 years	0	0.00
>4 years	30	100.00
Years of experience in green construction projects		
<2 years	0	0.00
2–3 years	10	33.33
3–4 years	9	30.00
>4 years	11	36.67
Respondent (total = 30)		
Designation		
Project director	3	10.00
Project manager	16	53.33
Project engineer	2	6.67
Safety manager	2	6.67
Quantity surveyor	7	23.33
Years of experience in conventional construction projects		
<2 years	0	0.00
2–3 years	2	6.67
3–4 years	9	30.00
>4 years	19	63.33

Continued

TABLE 10.2 Profile of Companies and Respondents—cont'd

Profile	Number	Percentage (%)
Years of experience in green construction projects		
<2 years	0	0.00
2–3 years	13	43.33
3–4 years	10	33.33
>4 years	7	23.33

TABLE 10.3 Profile of Projects

	Conventional Construction Projects (Total = 239)		Green Construction Projects (Total = 111)	
Project Profile	*Number*	*Percentage (%)*	*Number*	*Percentage (%)*
Type				
Commercial	52	21.76	38	34.23
Residential	101	42.26	31	27.93
Noncommercial	86	35.98	42	37.84
Nature				
New construction	184	76.99	93	83.79
Addition and alteration	55	23.01	18	16.21
Size				
<\$10M	48	20.08	12	10.81
\$10M to < \$20M	42	17.58	29	26.13
\$20M to < \$30M	58	24.28	7	6.31
\$30M to < \$40M	21	8.77	25	22.53
≥\$40M	70	29.29	38	34.22

A series of statistical test approaches was used to test and analyze the data collected from the questionnaire survey. First, the one sample *t*-test was conducted. This method is usually used to determine whether a sample comes from a population with a specific mean, which is not always known, but is sometimes hypothesized (Lam et al., 2011). For this chapter, the hypothesized value was established as 3, which represents the medium level of likelihood of occurrence and severity of each critical safety issue. Thus, by conducting the one sample *t*-test, it is able to verify whether each risk factor occurs significantly in or has a significant impact on green commercial construction projects. Then, the Spearman's rank correlation coefficient test was performed to examine the ranking agreement of critical safety issues between green and conventional construction projects. It is a method to compute a correlation between the ranks of scores of two groups, without considering the normality or equal variance of data (Zhao et al., 2013a; Hwang et al., 2015a). Finally, the Wilcoxon signed-rank test was conducted to check if a significant difference existed in the evaluations of the critical safety issues between green and conventional construction projects. The Wilcoxon signed-rank test is a nonparametric statistical hypothesis test used when comparing two matched samples to assess whether their population mean ranks differ, and this method has been widely used in previous construction engineering and management research when carrying out intergroup comparisons (Ameyaw et al., 2016).

10.3 ANALYSIS RESULTS AND DISCUSSIONS

10.3.1 Critical Safety Issues: Green vs Conventional Construction Projects

Because critical safety issues were evaluated in terms of both occurrence likelihood and severity, the final evaluation of each critical safety issue was determined by calculating the square root of the product of occurrence likelihood and severity. This approach has been adopted by Shen et al. (2001) and Ke et al. (2011) in their risk assessment studies, which are similar to the present study to some extent. Table 10.4 presents the evaluation of the respondents of safety issues in green and conventional construction projects, as well as the relevant statistical test results. The Spearman's rank correlation coefficient value of .915 (P-value = .000) indicates a strong correlation among the rankings of safety issues between green and conventional construction projects, and in particular, the two types of projects share the same top ten critical safety issues, namely, "fall from height," "electrocution," "being struck by falling objects," "contact with/by machinery," "fall on same level from slipping, tripping, and stumbling," "contact with hand-tools," "contact with heavy vehicles," "crush injuries," "fall on same level due to collision with, or pushing by, another person," and "moving/handling heavy loads."

TABLE 10.4 Comparison of Safety Issues Between Green and Conventional Construction Projects

	Green			Conventional			
Safety Issues ($\sqrt{\text{Likelihood} \times \text{Severity}}$)	*Rank*	*Mean*	P-*Value*	*Rank*	*Mean*	P-*Value*[a]	*P*-Value
Bad working positions (confined spaces)	13	2.95	.541	15	2.86	.110	.068
Building/structure collapse	23	2.54	.000[a]	23	2.54	.000[a]	1.000
Cave-ins	24	2.24	.000[a]	24	2.24	.000[a]	1.000
Collapse of tower crane	19	2.69	.003[a]	20	2.66	.001[a]	.317
Contact with/by machinery	4	3.37	.000[a]	4	3.38	.000[a]	.715
Contact with hand-tools	6	3.15	.072	7	3.14	.079	.655
Contact with heavy vehicles	7	3.14	.176	9	3.09	.379	.109
Crush injuries	8	3.14	.104	6	3.18	.039[a]	.492
Drowning	26	2.09	.000[a]	25	2.09	.000[a]	1.000
Electrocution	2	3.45	.000[a]	3	3.45	.000[a]	.317
Exposure to fire	11	3.06	.453	11	2.98	.800	.066
Exposure to hazardous substances	14	2.93	.538	16	2.83	.114	.039[b]
Exposure to heat/cold	22	2.6	.004[a]	22	2.59	.004[a]	.317
Fall on same level due to collision with, or pushing by, another person	9	3.12	.199	8	3.12	.199	1.000
Fall from height	1	3.86	.000[a]	1	3.86	.000[a]	1.000

Foreign matter in eye	20	2.66	.000[a]	21	2.6	.000[a]	.109
Inhalation	21	2.62	.000[a]	14	2.89	.299	.028[b]
Moving/handling heavy loads	10	3.1	.266	2	3.72	.000[a]	.000[b]
Respiratory failure	25	2.22	.000[a]	26	2.02	.000[a]	.001[b]
Fall on same level from slipping, tripping, and stumbling	5	3.3	.000[a]	5	3.31	.001[a]	.985
Being struck/crushed by vehicle	12	2.96	.629	18	2.8	.071	.147
Being struck against manually Operated tools	17	2.71	.003[a]	13	2.91	.463	.014[b]
Being struck by falling objects	3	3.39	.001[a]	10	3.09	.370	.005[b]
Being struck by moving objects	15	2.82	.029[a]	12	2.94	.594	.150
Traffic accidents on site	18	2.71	.005[a]	19	2.67	.001[a]	.701
Trapped/caught in/between objects	16	2.81	.035[a]	17	2.81	.035[a]	1.000
Spearman rank correlation coefficient: .915 (*P*-value = .000[c]).							

[a]*The one sample* t-*test result is significant at the level of 0.05 (two-tailed).*
[b]*The Wilcoxon signed-rank test result is significant at the level of 0.01 (two-tailed).*
[c]*The Spearman rank correlation is significant at the level of .01 (two-tailed).*

While rankings of safety issues among the two types of projects are generally similar, differences in the evaluations of critical safety issues between green and conventional construction projects are seen. The Wilcoxon signed-rank test results show that six out of 26 critical safety issues received significantly different evaluations between the two types of projects considering their corresponding P-values were less than the significance level at $\alpha = .05$. These safety issues are "exposure to hazardous substances," "inhalation," "moving/handling heavy loads," "respiratory failure," "being struck against manually operated tools," and "being struck by falling objects."

Among them, "inhalation" was regarded as a critical safety issue in conventional construction projects rather than in green construction projects, given that its mean value in conventional projects was statistically greater than the test value of 3 when performing the one sample t-test at the confidence level of 95%. This result can be explained by the fact that more environmentally friendly materials are used in green construction projects compared with those in conventional construction projects, which may reduce the possibility for frontline workers to inhale toxic gases (Cheng et al., 2015).

"Being struck by falling objects" was found to be a critical safety issue in both green and conventional construction projects, because its mean values in both types of project were statistically greater than or equal to the test value of the one sample t-test, namely, 3. In addition, this issue received a considerably higher evaluation of 3.39 (rank: third) in green construction projects than in conventional projects (mean value: 3.09, rank: tenth), which implies that it is more critical in green construction projects. This finding may be attributed to more prefabricated components being used in green construction projects to achieve a more environmentally responsible construction process, which inevitably requires more hoisting and lifting activities that result in more accidents caused by falling objects (Lu and Yuan, 2013).

Similarly, "moving/handling heavy loads" is also a critical safety issue to both green and conventional construction projects, given its mean evaluation and one sample t-test results. Unlike "being struck by falling objects," "moving/handling heavy loads" received a considerably lower mean evaluation of 3.1 (rank: tenth) in green construction projects than in conventional construction projects (mean value: 3.72, rank: second). This result indicates that the issue is more critical in conventional construction projects. This finding can be justified by the fact that heavy loads are more frequently moved and handled using construction machinery in green construction projects (Dowson et al., 2012), and thus workers employed on green construction projects face a lower risk in this regard.

The P-value of "exposure to hazardous substances" generated from the one sample t-test is $>.05$ in both green and conventional construction projects, which indicates that it is a critical safety issue in both types of projects. In addition, "exposure to hazardous substances" received a slightly higher evaluation of 2.93 in green construction projects than in conventional projects (mean value: 2.83). Superficially, such a result contradicts common knowledge because green construction projects are expected to adopt environmentally friendly materials. However, the respondents may be concerned that the materials adopted in green construction projects are typically newly developed, and thus lack sufficient test in practice (Reijnders and Van Roekel, 1999), which will probably be more potentially hazardous to the health of frontline workers.

"Being struck against manually operated tools" was found to be a critical safety issue in conventional construction projects rather than green construction projects, given its mean value in the latter was statistically less than the test value of the one sample *t*-test, namely, 3. This finding may be ascribed to the fact that manually operated tools, such as saws, drills, hammers, and grinders are more widely used in conventional construction projects, so that frontline workers working on these projects are comparatively more easily hurt by these tools. (Aghazadeh and Mital, 1987).

Because its mean values in both green and conventional construction projects were statistically less than the test value of the one sample *t*-test, "respiratory failure" tends not to be a critical safety issue in either type of project, and was thus excluded from further analysis.

10.3.2 Safety Performance: Green vs Conventional Construction Projects

Inspired by Cameron et al. (2008), this chapter uses accident rates as the safety performance metric of green and conventional construction projects. This metric was calculated by dividing the number of accidents by the number of projects, as shown in Table 10.5. The overall accident rate in green construction projects is 0.37, which is higher than that of conventional construction projects (0.31). Moreover, the accident rates in green commercial, residential, new construction, addition and alteration, and construction projects with an investment of over SGD 20 million, are all greater than those in conventional construction projects. These results indicate that green construction projects generally suffer from a worse safety performance than conventional construction projects. This finding is not surprising because the number of green construction projects in Singapore has achieved rapid growth in recent years, which inevitably increases accident rates in these projects. Nonetheless, a few exceptions can be noted in projects with an investment of less than SGD 20 million and in non-commercial projects, in which the accident rates in green construction projects are equal to or even lower than those in conventional construction projects. However, these exceptions are merely discrete results that cannot reverse the general situation.

TABLE 10.5 Comparison of Safety Performance Between Green and Conventional Construction Projects

Project Profile		No. of Conventional Construction Projects Performed (a)	No. of Accidents (b)	Accident Rate (b)/(a)	No. of Green Construction Projects Performed (c)	No. of Accidents (d)	Accident Rate (d)/(c)
Project type	Commercial	52	31	0.60	38	23	0.61
	Residential	101	26	0.26	31	10	0.32
	Noncommercial	86	17	0.20	42	8	0.19
Project nature	New construction	184	54	0.29	93	33	0.36
	Addition and alteration	55	20	0.36	18	8	0.44
Project size	<$10M	48	4	0.08	12	1	0.08
	$10M to < $20M	42	7	0.17	29	5	0.17
	$20M to < $30M	58	12	0.20	7	2	0.28
	$30M to < $40M	21	10	0.47	25	13	0.52
	≥$40M	70	31	0.43	38	20	0.53
Total		239	74	0.31	111	41	0.37

10.3.3 Solutions for Safety Performance Improvement of Green Construction Projects

Let us turn to the ideas of the respondents regarding solutions that may help improve the safety performance in green construction projects. Ten possible solutions were listed in the survey and they were evaluated using two different rating systems. First, a five-point Likert rating system was adopted to gauge the effectiveness of each proposed solution. Simultaneously, a preference rating system of the top five selected solutions was also utilized for all the proposed solutions. Table 10.6 presents the rankings of the solutions based on the two rating systems. Notably, the two types of ratings recommended the same top five solutions, which are discussed as follows.

Ensure that all workers undergo adequate workplace safety health training to be proficient in his/her trade. This solution was ranked first in both the Likert rating and the frequencies of the top five selected solutions. Kaskutas et al. (2013) claimed that safety training could not only help raise safety awareness, but could also improve worker safety knowledge, standardize safety behavior, encourage mentoring of fellow workers, and cultivate a positive safety culture. Numerous studies have also emphasized that safety training is a major solution for sustaining and improving safety performance in construction projects (Demirkesen and Arditi, 2015; Choudhry et al., 2008; Tam et al., 2004; Jaselskis and Recarte Suazo, 1994).

Ensure the adequate provision of personal protective equipment (PPE) and enforce adherence to safe work practices and procedures on construction sites at all times. This solution was ranked second in both the Likert rating and the frequencies of the top five selected solutions. The main purpose of wearing PPE is to protect workers from exposure to dangers in their working environment. However, factors such as garment wearability, comfort, fit, and style, frequently make frontline workers reluctant to wear PPE (McPherson, 2008). Moreover, providing PPE without enforcement to ensure safe work practices often leads to unsafe operating procedures and increased workplace injury rates (Huang and Hinze, 2003; Williams Jr. et al., 2010; Contreras and Buchanan, 2012). Therefore, ensuring that PPE is used properly is crucial to reduce accident rates.

Implement the design for safety (DFS) concept as a core syllabus in preemployment professional courses for architecture, engineering, project management, and real estate at the tertiary level. This solution was ranked third in the frequencies of the top five selected solutions and fifth in the Likert rating. Prevention through design is an effective measure to improve safety performance (Zhou et al., 2015), because construction safety risk can be mitigated through design decisions made prior to working in the construction sector (Gangolells et al., 2010). Zhou et al. (2012) found that a close working relationship between the builders and designers, as well as knowledge of construction, is crucial to the occupational health and safety of construction workers. Therefore, the DFS concept should be a part of the core syllabus of construction-

TABLE 10.6 Solutions for Improving Green Construction Project Safety Performance

No.	Solutions	Mean	Rank	Top Five Selected	Frequency (%)	Rank
1	Ensure that all workers will undergo adequate workplace safety health training to be proficient at his/her trade.	4.6	1	26	86.67	1
2	Ensure the adequate provision of personal protective equipment (PPE) and enforce adherence to safe work practices and procedures on construction sites at all times.	4.53	2	25	83.33	2
3	Conduct more regular audits to assess the quality of the safety, health, and management system of a construction firm.	4.33	3	17	56.67	5
4	Ensure that the safety policy of a company is in place to obtain full commitment from the top management all the way down the hierarchy to guarantee the understanding and compliance of the workers on site.	4.30	4	15	50.00	4
5	Implement the design for safety (DFS) concept as a core syllabus in preemployment professional courses for architecture, engineering, project management, and real estate at tertiary level.	4.17	5	18	60.00	3
6	Promote safety culture through educating and recognizing safety efforts. For e.g., safest track record with lowest injury rate.	4.10	6	11	36.67	6
7	Ensure that all plants, machineries, tools and equipment that are used on site are serviced according to the manufacturers' recommendation.	4.10	7	7	23.33	8
8	Prerequisite to have a fall protection plan developed and implemented on construction site.	4.10	8	12	40.00	7
9	Instill higher penalty/fine for poor safety management.	4.00	9	8	26.67	9
10	Prerequisite for all stakeholders at management level to attend safety courses to be well versed with risk management and soft skills such as communications and supervisory skills.	3.83	10	6	20.00	10

related professional courses at the tertiary level to address and mitigate the issue of poor safety performance at the early design phase.

Conduct regular audits to assess the quality of the safety, health, and management system of a construction firm. This solution was ranked third in the Likert rating and fifth in the frequencies of the top five selected solutions. A safety, health, and management system plays a vital role in ensuring the safety performance of a construction firm (Goh and Chua, 2013). However, the regular auditing of such a system is even more crucial because such a process not only determines whether a construction firm is complying with its safety standards, policies, and procedures, but can also examine the effectiveness of the system. Through regular audits, the deficiencies of a system can be determined and the positive achievements can be identified and reinforced.

Ensure that the safety policy of a company is in place to obtain full commitment from the top management all the way down the hierarchy to guarantee the understanding and compliance of the workers on site. This solution was ranked fourth in both the Likert ranking and the frequencies of the top five selected solutions. Construction organizations have large, complex, and decentralized structures, inevitably leading the practitioners at different levels and among various subunits to have different levels of safety awareness (Zhang et al., 2015). To improve safety performance, safety should be a value and a culture that requires the full commitment of an organization at all its levels (Abudayyeh et al., 2006). Correspondingly, safety policies should be established and implemented to embody the value of safety in practice.

10.4 SUMMARY

While considerable research effort has been directed to the area of green construction, little research has been conducted to examine safety problems in green construction projects. To bridge this gap, a questionnaire survey was administered to 30 Singapore-based construction companies to investigate the safety performance and safety issues in green construction projects. Results showed that, compared to conventional construction projects, green construction projects were affected by poor safety performance. Results also showed that, although the two types of project share an identical list of top ten critical safety issues, namely, "fall from height," "electrocution," "being struck by falling objects," "contact with/by machinery," "fall on same level from slipping, tripping, and stumbling," "contact with hand-tools," "contact with heavy vehicles," "crush injuries," "fall on same level due to collision with, or pushing by, another person," and "moving/handling heavy loads," six of them were perceived significantly differently by project type. Specifically, "exposure to hazardous substances," "respiratory failure," and "being struck by falling objects" were found to be more critical in green construction projects than in conventional ones; whereas "inhalation," "moving/handling heavy loads," and "being struck against manually operated tools," were found more critical in conventional construction projects than in their green counterparts. Lastly, the top five effective solutions that may help improve the safety performance in green construction projects, based on the feedback collected from the survey, were discussed in this chapter.

This is the first time that the safety performance, safety issues and the relevant improvement solutions in green construction projects have been systematically investigated, and thus it indeed contributes to the body of knowledge. In addition, the comprehensive framework of safety issues identified by this chapter can also serve as a solid basis for further studies in this domain, which benefits the academic researchers also. In the meantime, the findings from this chapter are helpful to industry practitioners because they alert them to the serious safety problems in green construction projects. Moreover, the findings from this chapter can also assist the practitioners in identifying the critical safety issues they may encounter during the implementation of green construction projects and in developing effective measures to cope with those safety issues.

Chapter 11

Rework Performance and Improvement of Green Construction Projects

11.1 INTRODUCTION

Defined as the unnecessary effort of re-doing a process or activity that was incorrectly implemented the first time (Hwang et al., 2014; Love and Edwards, 2004), rework has been widely recognized as a significant concern to the construction industry. Rework may originate from a large number of sources including construction changes, design errors and omissions, or coordination issues among contracting parties at the site (Palaneeswaran et al., 2008; Hwang et al., 2014) and may eventually lead to cost and schedule overruns (Hwang et al., 2009). According to the Construction Industry Institute (2005), costs caused by rework amounted to 5% of total construction expenditures in the US construction industry. In the Australian construction industry, main direct and indirect rework costs were found to be 6.4% and 5.6% of the original contract value, respectively (Love, 2002). In Singapore, rework has been accused of contributing to an average of 25% of the construction schedule growth (Hwang and Yang, 2014). These findings reveal that rework is a significant issue affecting project performance in the construction industry.

Green construction projects also suffer from rework (Chandramohan et al., 2012). This is because green construction projects always tend to use innovative materials and complicated technologies to reinforce their green performance and these green materials and technologies may be unstable and cause contingencies to projects, thus necessitating rework (Kang et al., 2013). However, until now, very little research effort has been made to investigate the rework issues in green construction projects. Therefore, using green construction projects in Singapore as a background, the aims of this chapter are: (1) assess the status quo of rework in green construction projects, (2) identify and evaluate the rework factors in green construction projects, (3) compare their criticalities with their conventional counterparts, and (4) propose feasible solutions to reduce rework in green construction projects. This chapter contributes to the body of knowledge by exploring the rework problem in green construction projects.

Performance and Improvement of Green Construction Projects. https://doi.org/10.1016/B978-0-12-815483-0.00011-9

Also, this chapter is beneficial to the industry as it can provide practitioners with a comprehensive picture of reworks in green construction projects and a set of feasible solutions as well.

11.2 METHOD AND DATA COLLECTION

To achieve the research aims outlined above, a comprehensive literature review to identify the various rework factors and solutions in green construction projects was conducted. As the research of rework related to green construction projects is limited, this chapter expands on the literature search scope to incorporate rework studies on conventional construction projects. Based on the literature review, 8 rework factors and 11 solutions were identified, as shown in Tables 11.1 and 11.2. The identified rework factors and solutions formed a self-explanatory questionnaire that consisted of six sections. Sections A and B gathered respondents' and their companies' demographics, respectively. Section C solicited the status quo of rework within the green and conventional construction projects undertaken by the respondents' companies. Sections D and E asked respondents to evaluate the likelihood and impact of rework factors within green and conventional construction projects, using two five-point Likert rating scales (1 = least likely/least significant, 2 = less likely/less significant, 3 = neutral/neutral, 4 = more likely/more significant, and 5 = most likely/most significant). Section F asked respondents to assess the effectiveness of the identified solutions in the context of green construction projects.

The questionnaire was targeted at industry experts with experience in both green and conventional construction projects in Singapore, and a total of 127 questionnaires were disseminated to BCA certified companies via electronic mail. The data collection effort produced 30 completed questionnaires from 30 different companies, yielding a response rate of 23.6%, which is consistent with the norm of 20%–30% in most surveys in the construction industry (Akintoye, 2000). Tables 11.3 and 11.4 profile the respondents, and the companies and projects sampled in the survey, respectively. It is noted that 90% of companies and >50% of respondents had more than 10 years of experience in conventional construction projects, and that 70% of companies and 60% of respondents had three or more years of experience in green construction projects, indicating that respondents were experienced enough to address the research questions of the survey.

A series of statistical methods were used to analyze the data collected from the questionnaire. The Chi-squared test was conducted to check if the occurrence of rework was associated with the sustainable nature of the projects. This method was selected as it is a common approach to check for significant differences between the expected frequencies and the observed frequencies in one or more categories (Uher and Brand, 2008). The one sample *t*-test was performed to verify whether each rework factor occurred significantly in, or had a

TABLE 11.1 Rework Factors Identified From Literature

	Source												
Cause of Rework	*A*	*B*	*C*	*D*	*E*	*F*	*G*	*H*	*I*	*J*	*K*	*L*	*M*
Owner change			√	√	√	√	√			√		√	
Design error/omission	√	√	√	√	√	√	√	√	√		√		√
Design change	√		√	√	√	√	√	√			√	√	√
Contractor's error/omission			√	√			√		√		√		√
Constructor change			√	√			√					√	√
Vendor error/omission				√	√		√		√		√		
Vendor change				√	√		√					√	
Transportation error					√		√		√		√		

Note: A, Love et al. (1999); B, Love et al. (2000); C, Love and Li (2000); D, Josephson et al. (2002); E, Love and Smith (2003); F, Love and Edwards (2004); G, Hwang et al. (2009); H, Love et al. (2011); I, Love et al. (2014); J, Hwang et al. (2014); K, Hwang and Yang (2014); L, Forcada et al. (2014); and M, Ekambaram et al. (2014).

TABLE 11.2 Solutions Identified From Literature Review

	Source						
Solution to Reduce Rework	*A*	*B*	*C*	*D*	*E*	*F*	*G*
Use communication tools for design coordination	√	√			√		
Ensure the correctness of design	√	√					
Ensure the compatibility of design and specifications	√	√					
Establish an effective channel for communication	√	√			√	√	
Ensure a sound change management		√		√	√		
Design review and verification	√	√					
Interorganizational collaboration and learning	√					√	√
Corrective action planning		√	√	√			
Front-end planning		√	√				
Enhanced access to information	√	√			√		
Rework-tracking system		√		√			

Note: A, Love and Li (2000); B, Josephson et al. (2002); C, Love and Smith (2003); D, Zhang et al. (2012); E, Hwang and Yang (2014); F, Love et al. (2014); and G = Love et al. (2015b).

TABLE 11.3 Demographics of Respondents and Companies

Profile	Number	Percentage (%)
Respondent (N=30)		
Occupation		
Contractors	13	43.33
Architects	5	16.67
Quantity surveyors	3	10.00
Consultants	9	30.00
Years of experience in conventional building construction projects		
<10 years	13	43.33
10–20 years	14	46.67
20–30 years	2	6.67
>30 years	1	3.33
Years of experience in green building construction projects		
<3 years	12	40.00
3–5 years	12	40.00
5–10 years	6	20.00
>10 years	0	0.00
Number of conventional building construction projects involved in past 5 years		
<3	2	6.67
3–5	10	33.33
5–10	13	43.33
>10	5	16.67
Number of green building construction projects involved in past 5 years		
<3	16	53.33
3–5	8	26.67
5–10	6	20.00
>10	0	0.00
Company (N=30)		
Type		
Architecture	6	20.00
Developer	7	23.33

Continued

TABLE 11.3 Demographics of Respondents and Companies—cont'd

Profile	Number	Percentage (%)
Quantity surveyor	4	13.33
Contractor	13	43.33
Years of experience in conventional building construction projects		
<10 years	3	10.00
10–20 years	25	83.33
20–30 years	2	6.67
>30 years	0	0.00
Years of experience in green building construction projects		
<3 years	9	30.00
3–5 years	15	50.00
5–10 years	6	20.00
>10 years	0	0.00

TABLE 11.4 Demographics of Building Construction Projects Undertaken By Respondents' Companies

	Traditional Building Projects (Total = 210)		Green Building Projects (Total = 102)	
Project Profile	*Number*	*Percentage (%)*	*Number*	*Percentage (%)*
Type				
Residential	50	23.81	15	14.71
Industrial	19	9.05	13	12.75
Office	128	60.95	64	62.75
Retail	13	6.19	10	9.80
Size (million)				
<SGD 15	99	47.14	37	36.27
SGD 15–SGD 50	76	36.19	32	31.37
SGD 50–SGD 100	16	7.62	19	18.63
SGD 100<	19	9.05	14	13.73

significant impact, on green and conventional building construction projects. This method is usually used to examine the mean difference between the sample and the known value of the population mean (Lam et al., 2011). The Wilcoxon signed-rank test was adopted to check the differences among the evaluations of rework factors between green and conventional building construction projects. This method is widely used in comparing two matched or related samples and determining whether there is a significant difference between those two samples (Ameyaw et al., 2016). The Spearman's rank correlation was performed to examine the ranking agreement of rework factors between green and conventional building construction projects. This is a widely used method to compute the correlation between the ranks of scores of two groups (Hwang et al., 2015a).

11.3 ANALYSIS RESULTS AND DISCUSSIONS

11.3.1 Rework Performance: Green vs Conventional Construction Projects

Table 11.5 shows the status of rework in green and conventional construction projects involved in the survey. To check if there is a relationship between the sustainable nature of a project and its rework status, a Chi-squared test was conducted. Data collected from the questionnaire were input into SPSS Statistics 17.0 software to perform the test, and the test reported a *P*-value of .014, which was less than the significance level of 0.05. This result indicates a significant relationship between the projects' sustainable nature and the rework status.

TABLE 11.5 Rework Status of Projects Involved in the Survey

	Conventional Construction Projects (Total = 210)		Green Construction Projects (Total = 102)	
Project Rework Status	*Number*	*Percentage (%)*	*Number*	*Percentage (%)*
With rework	92	43.81	30	29.41
Cost overrun	21	22.83	2	6.67
Schedule delay	40	43.48	7	23.33
Cost overrun and project delay	31	33.70	21	70.00
Without rework	118	56.19	72	70.59
P-value of Chi-squared test = .014[a]				

[a]*The Chi-squared test result is significant at the level of 0.05 (two-tailed).*

Moreover, Table 11.5 shows that 29.41% of green construction projects experienced rework, while conventional construction projects experienced 43.81%. Such a significant discrepancy reveals that green construction projects enjoyed a relatively lower incidence of rework, a result beneficial for the further promotion of green construction projects.

Cost overrun and schedule delay are two inevitable consequences of rework (Hwang and Leong, 2013; Love, 2002; Love et al., 2010) and they were also examined in frequency within the two types of projects in the questionnaire. Results in Table 11.5 show that, among the 92 conventional construction projects where rework had occurred, 21 (23%) suffered cost overrun, 40 (43%) suffered schedule delay, and 31 (34%) suffered both. By contrast, among 30 green construction projects in which rework had occurred, two (7%) suffered cost overrun, seven (23%) suffered schedule delay, and 21 (70%) suffered both. Obviously, green construction projects are more vulnerable to the adverse impacts of rework than conventional construction projects. This may be attributed to the fact that green construction projects are more complex in their design, construction technologies, and materials, and hence, more serious consequences would be incurred if rework occurs in such projects (Yang et al., 2016; Shi et al., 2016).

11.3.2 Criticalities of Rework Factors: Green vs Conventional Construction Projects

Table 11.6 presents the evaluation results of rework factors in green and conventional construction projects. According to the evaluations in likelihood, the top four significant rework factors in green construction projects were "owner change," "design change," "design error/omission," and "contractor's error/omission." This was further proven by the one sample *t*-test (test value = 3) because the mean scores for these factors were statistically greater or equal to the test value at the 95% confidence interval. By contrast, the other four rework factors occurred insignificantly in green construction projects as their likelihood evaluations were less than the test value at the 95% confidence interval. In addition, the Spearman rank correlation coefficient for the two types of projects was .976 (P-value = .000), which suggests that the rankings of rework factors in likelihood evaluation were highly associated at the 99% confidence interval.

"Owner change" received the highest value in the likelihood evaluation and thus was regarded as the rework factor that occurred most frequently in green construction projects. This is probably because sustainability was an exclusive goal set up by the owner for green construction projects (Li et al., 2011). The owners might propose as many changes as they can to achieve the goals and maximize the environmental benefits, and this would inevitably lead to more reworks. In addition, the survey results revealed that "owner change" scored similarly between green and conventional construction projects, suggesting that the rework factor was of similar likelihood of occurrence between the two types of projects.

TABLE 11.6 Evaluations of Rework Factors in Green and Conventional Construction Projects

Rework Factor	Likelihood							Impact							Criticality						
	Green Projects			*Conventional Projects*				*Green Projects*			*Conventional Projects*				*Green Projects*			*Conventional Projects*			
	Rank	*Mean*	P-*Value*	*Rank*	*Mean*	P-*Value*	P-*Value*	*Rank*	*Mean*	P-*Value*	*Rank*	*Mean*	P-*Value*	P-*Value*	*Rank*	*Mean*	P-*Value*	*Rank*	*Mean*	P-*Value*	P-*Value*
Owner change	1	4.43	.000[a]	1	4.47	.000[a]	.317	1	4.37	.000[a]	1	4.37	.000[a]	1.000	1	19.73	.000[a]	1	19.93	.000[a]	.285
Design error/ omission	3	3.17	.305	3	3.07	.625	.366	3	3.10	.522	3	2.97	.823	.046[b]	3	10.23	.173	3	9.43	.578	.001[b]
Design change	2	4.00	.000[a]	2	3.57	.000[a]	.000[b]	2	3.73	.000[a]	2	3.43	.000[a]	.003[b]	2	14.90	.000[a]	2	12.30	.000[a]	.041[b]
Contractor's error/ omission	4	2.80	.110	4	2.63	.009[a]	.025[b]	4	3.00	1.000	4	2.87	.403	.046[b]	4	8.60	.479	4	7.77	.031[a]	.010[b]
Constructor change	5	2.07	.000[a]	5	1.83	.000[a]	.059	5	1.90	.000[a]	5	1.93	.000[a]	.317	5	4.23	.000[a]	5	3.80	.000[a]	.074
Vendor error/ omission	6	1.77	.000[a]	6	1.70	.001[a]	.317	5	1.90	.000[a]	6	1.87	.000[a]	.317	6	3.83	.000[a]	6	3.50	.000[a]	.317
Vendor change	8	1.53	.000[a]	7	1.53	.000[a]	1.000	7	1.73	.000[a]	7	1.73	.000[a]	1.000	7	3.40	.000[a]	7	3.00	.000[a]	.655
Transportation error	7	1.63	.000[a]	8	1.47	.000[a]	.025[b]	8	1.57	.000[a]	8	1.50	.000[a]	.157	8	3.00	.000[a]	8	2.97	.000[a]	.024[a]
Spearman rank correlation coefficient	.976						.000[c]	.994						.000[c]	1.00						.000[c]

[a]*The one sample* t-*test result is significant at the level of 0.05 (two-tailed).*
[b]*The Wilcoxon signed-rank test result is significant at the level of 0.01 (two-tailed).*
[c]*The Spearman rank correlation is significant at the level of .01 (two-tailed).*

"Design change" was ranked second in the likelihood evaluation in green construction projects. As is well-known, prominent characteristics of green construction projects are their unique and innovative designs. However, a concomitant issue of these unique and innovative designs is that they might be less commonly used in practice and thus sufficient application experience may be lacking (Kang et al., 2013). Under such circumstances, "design change" occurs inevitably and finally results in rework. Moreover, based on the result of the Wilcoxon signed-rank test, the mean score of "design change" in green construction projects (4.00) was statistically higher than that in conventional construction projects (3.57) at the 99% confidence level, suggesting the rework factor occurred more frequently in green construction projects than in conventional construction projects. This result also echoes the foregoing justification that uncertainties in innovative designs in green construction projects were apt to lead to design changes and thus, eventually, reworks.

"Design error/omission" was ranked third in the likelihood evaluation in green construction projects. In fact, numerous previous studies have remarked that "design error/omission" was a significant issue that leads to rework in conventional construction projects (Love et al., 2000; Josephson et al., 2002; Love and Smith, 2003; Hwang et al., 2009), and it has been further demonstrated that this factor is also a significant contributor to rework in green construction projects. However, the mean score of "design error/omission" (3.17) was considerably lower than that of "design change" (4.00), the second most frequent rework factor, which indicated that the former occurred significantly less frequently than the latter. Such a discrepancy also coincides with the commonsense observation that error/omission occurs sometimes rather than always.

"Contractor's error/omission" received the fourth ranking in the likelihood evaluation in green construction projects, with a mean evaluation of 2.80. The occurrence of "contractor's error/omission" is mainly up to the familiarity of the contractor with the construction technologies and methods applied during project implementation, and green construction projects always tend to use some new and innovative technologies and methods that might be unfamiliar to contractors at times (Kang et al., 2013). Thus, errors and omissions of contractor workers might be increased by their improper activities and eventually result in rework. In addition, this result implies indirectly that experienced contractors with expertise in green construction in the Singaporean construction industry are currently scarce.

According to Table 11.6, rework factors that have significant impacts on green construction projects are "owner change," "design change," "design error/omission," and "contractor's error/omission," because their mean scores were statistically greater or equal to the test value of the one sample t-test, namely, 3 at the 95% confidence interval. Meanwhile, the calculated Spearman rank coefficient for the rankings of impact evaluations of rework factors within green and conventional construction projects is .994 (P-value = .000), suggesting the rankings within the two types of projects are highly associated.

"Owner change" was ranked first with the highest value of 4.37 in the impact evaluation, suggesting it was the most impactful rework factor in green construction projects. Such a result was unsurprising because the owners occupy the predominant position in the current construction industry, and their change orders are always inarguable and therefore impact projects significantly. Similar conclusions were also achieved by previous studies that "owner change" was the most impactful factor on the project performance in conventional construction projects (Josephson et al., 2002; Love and Smith, 2003; Hwang et al., 2009, 2014).

"Design change" received the second ranking and a high value of 3.73 in impact evaluation, indicating that it was also an impactful rework factor in green construction projects. This result echoes Josephson et al. (2002) that any change related to design would definitely cause concomitant consequences and would probably result in rework. In addition, the result of the Wilcoxon signed-rank Test revealed that the mean score of "design change" in green construction projects (3.73) was statistically higher than that in conventional construction projects (3.43) at the 99% confidence interval, which indicated that the rework factor was more impactful in green construction projects. This could be attributed to the fact that design was more critical to green construction projects than conventional construction projects (Li et al., 2011), and thus variations of design in green construction projects would cause worse consequences.

"Design error/omission" and "Contractor's error/omission" were the third and fourth most impactful rework factors in green construction projects, respectively. The mean values of these two rework factors (3.10 and 3.00) were considerably lower than those of the top two rework factors (4.37 and 3.73), which indicated that the impacts of "design error/omission" and "contractor's error/omission" on green construction projects were significantly less than the top two rework factors. Nonetheless, the results of mean evaluation and the Wilcoxon Signed-rank Test showed that the impacts of "design error/omission" and "contractor's error/omission" on green construction projects were greater than those on conventional construction projects, which implies that special attention to these two rework factors in green construction projects is required.

To reflect the rework factors more comprehensively, the concept of criticality was adopted, as inspired by Shen et al. (2001). The criticality of each rework factor was determined by calculating the product of its likelihood and impact evaluations, and the relevant calculation results, as well as its statistical test results, are presented in Table 11.6. Results showed that the most critical rework factors in green construction projects were "owner change," "design change," "design error/omission," and "contractor's error/omission," because their mean scores were statistically greater or equal to the test value of the one sample t-test, namely, 9, the product of the median values of likelihood and impact rating scales. In addition, the result of the Wilcoxon signed-rank test revealed that the criticality evaluations of "design change," "design error/omission," and "contractor's error/omission" in green construction projects were statistically higher than those in conventional construction projects at the 99% confidence interval, indicating that these three rework factors are more critical to green construction projects than conventional construction projects. Such results are generally in line with the likelihood and impact evaluation results of rework factors in green construction projects, which further proves that the four rework factors were of particular importance to the rework problem in green construction projects.

11.3.3 Solutions to Reduce Rework in Green Construction Projects

Eleven solutions to reduce rework in green construction projects were recommended in this chapter, and they were also assessed in the questionnaire survey. Two different assessment approaches were employed. First, a five-point Likert rating scale (1 = least effective and 5 = most effective) was utilized to measure the effectiveness of solutions. Second, a ranking preference rating system was adopted to request respondents to rank the whole 11 solutions from 1 (the most effective) to 11 (the least effective). To highlight the most effective solutions, evaluation results generated from the preference rating approach were reported in the frequency of the top five selected solutions. Table 11.7 illustrates the evaluation results. For the results generated from the Likert rating, a one sample t-test was conducted with a hypothesized value of 3 to check the effectiveness of each recommended solution. The test results revealed that, "interorganizational collaboration and learning," "enhanced access to information," and "rework-tracking system" were statistically insignificant because their mean values and relevant P-values were <3 at the 95% confidence interval; while the rest of the eight solutions were found to be statistically significant. In addition, the calculated Spearman rank correlation coefficient value for the two rating systems was .878 with a P-value of .000, suggesting that the two rankings of solutions were highly associated. Meanwhile, it was noteworthy that the two rating approaches reached the same top five effective solutions, namely, "use communication tools for design coordination," "ensure the effectiveness of design," "ensure the compatibility of design and specifications," "establish an effective channel for communication," and "ensure a sound change management."

The results in Table 11.7 show that the top three solutions generated from the Likert rating were all design-related solutions, and in particular, two of them, namely, "ensure the correctness of design" and "ensure the compatibility of design and specifications," were tied for first in the top five selected rating. This result indicates that design is crucial to the reduction of reworks in green construction projects. "Ensure the correctness of design" was ranked first in the top five selected rating and second in the Likert rating. This could be attributed to the fact that construction of a facility is highly dependent on design, and any error or omission in the design documents could affect the construction process

TABLE 11.7 Solutions to Reduce Reworks in Green Construction Projects

	Likert Rating Scale			Top Five Selected	
Solutions	*Rank*	*Mean*	*P-Value*	*Rank*	*Frequency*
Use communication tools for design coordination	1	3.60	.000[a]	4	17
Ensure the correctness of design	2	3.53	.003[a]	1	23
Ensure the compatibility of design and specifications	3	3.33	.030[a]	1	23
Establish an effective channel for communication	4	3.30	.037[a]	3	21
Ensure a sound change management	5	3.03	.879	5	16
Design review and verification	6	2.70	.095	8	9
Interorganizational collaboration and learning	7	2.53	.017[a]	9	8
Corrective action planning	7	2.53	.100	6	13
Front-end planning	9	2.43	.088	6	13
Enhanced access to information	10	2.33	.000[a]	10	3
Rework-tracking system	11	1.60	.000[a]	10	3
Spearman rank correlation coefficient = .878 (*P*-value = .000[b])					

[a]*The one sample t-test result is significant at the level of 0.05 (two-tailed).*
[b]*The Spearman rank correlation coefficient is significant at the level of .01 (two-tailed).*

and eventually result in rework (Hwang and Yang, 2014). To avoid design error or omission, the respondents further suggested that the drawings could undergo a cross-check before they are delivered. "Use communication tools for design coordination" and "ensure the compatibility of design and specifications" received the first and third ranks in the Likert rating, respectively. Design work of a facility is comprised of several different branches, such as architectural, structural and mechatronic design. Normally each branch has its own design protocol and standard to follow during the design process (Cheng et al., 2013). However, conflicts and clashes might occur when designs from different branches are integrated, which would lead eventually to rework. To address this issue, the application of design coordination tools was suggested. Design coordination tools refer to computer-aided design modeling instruments, such as AutoCAD and REVIT, which can facilitate information transfer and coordination among stakeholders efficiently (Zaneldin et al., 2001). By utilizing these tools, project stakeholders can gain a better understanding of the work to be performed by other parties, and design issues, such as conflicts and design errors, may be identified and rectified in the early stages of the project, thus minimizing rework. Meanwhile, the use of design coordination tools can enhance the compatibility of design and specifications, also reducing the likelihood of occurrence of rework in green construction projects.

"Establish an effective channel for communication" was ranked fourth in the Likert rating scale and third in the top five selected rating. Communications among different stakeholders is one of the significant challenges in undertaking green construction projects (Hwang and Ng, 2013). Compared with conventional construction projects, green construction projects involve more complex technologies and designs that require more communication among project stakeholders (Kang et al., 2013). Therefore, it is extremely important to establish an effective channel to improve the communication among the various project stakeholders. The respondents also mentioned that such channels could include periodic project meetings and integrated construction management systems.

"Ensure a sound change management" received the fifth ranking in both the Likert and the top five selected ratings. Change orders are one of the major sources of rework in construction projects (Love and Li, 2000; Hwang and Yang, 2014). They can arise from various origins, such as owner-induced enhancements, design errors and omissions, and contract omissions. A sound change management is thus crucial to the reduction of rework in construction projects. To mitigate reworks caused by change orders, two specific measures were suggested by the respondents. First, a high-quality design with the least possible design errors and omissions should be generated and fully communicated with the owner to make sure that the owner's intent and requirements for the project are embodied thoroughly in the design. Second, a detailed change order database was strongly recommended. The database would collect various types of change orders in terms of their sources, nature, and final treatments. It is believed that such a database could render valuable reference of change management for future projects.

11.4 SUMMARY

A detailed investigation of rework issues in green construction projects was performed in this chapter. Results showed that although the incidence of rework in green construction projects was lower than that of conventional construction projects, the adverse impacts of rework on green construction projects were higher than those on conventional projects. This chapter also unveiled that the top four critical rework factors in green construction projects were "owner change," "design change," "design error/omission," and "contractor's error/omission," and that these four rework factors were generally more critical to green construction projects than conventional projects. Finally, five practical solutions were recommended in this chapter to reduce reworks in green construction projects, namely "use communication tools for design coordination," "ensure the correctness of design," "ensure the compatibility of design and specifications," "establish an effective channel for communication," and "ensure a sound change management."

Chapter 12

Productivity Performance and Improvement of Green Construction Projects

12.1 INTRODUCTION

Productivity is generally used to describe the relationship between output and the associated inputs used to generate that output, assessing the effectiveness and efficiency of resource usage in achieving the output (SPRING Singapore, 2011; Construction Industry Institute (CII), 2010). It is crucial for both countries and organizations to achieve and maintain long-term competitiveness and profitability (SPRING Singapore, 2011). For instance, prior to the 1997 Asian financial crisis, the increase in productivity contributed to 60% of Singapore's economic growth and because of this Singapore was once labeled as one of the Four Asian Tigers (Kiat, 1991).

As mentioned earlier in Chapter 1, Singapore has established an ambitious target to have 80% of its buildings go green by 2030. Construction productivity can be a basic and determining factor to secure the achievement of this target. However, the declining productivity of the construction industry in Singapore poses a great question mark to the achievement of this target. Labor productivity, which is a crucial productivity index of the Singapore construction industry, continuously declined from 0.04 in 2010 to −0.027 in 2013. Worse still, the reported productivity values for the first and second quarter in 2014 were −0.007 and −0.020, respectively (Department of Statistics Singapore, 2014). In light of this situation, BCA implemented two rounds of a Construction Productivity Roadmap in 2011 and 2015, respectively, to improve the productivity in Singapore over the long term (BCA, 2011, 2015c). With the endorsement of the National Productivity and Continuing Education Council, the goal of the first Roadmap was to have a competent construction sector that is advanced and supported by a skilled workforce by 2020 (BCA, 2011), while the second Roadmap aimed to equip the construction industry with high technologies (BCA, 2015c).

Furthermore, the low productivity of the development of green buildings also poses a great challenge to the green target because a low construction productivity generally leads to cost and schedule problems (Doloi et al., 2012;

Performance and Improvement of Green Construction Projects. https://doi.org/10.1016/B978-0-12-815483-0.00012-0

Gündüz et al., 2013). First, the capital costs of green building developments are perceived to be higher than those of conventional buildings, leading to less development of the green construction industry (Dwaikat and Ali, 2016). As explained earlier in previous chapters, the design and construction of green buildings is different from that of conventional buildings, which usually leads to high design and construction costs (Robichaud and Anantatmula, 2011; Chan et al., 2009). In order to motivate the industry, firms are recommended to change their business models to maximize their profits from green constructions (Mokhlesian and Holmén, 2012). In addition, the cost overrun is another hindrance, which could be caused by the many requirements to achieve a green certificate, as well as the technical difficulties during a construction process resulting from unfamiliarity with green technologies (Zhang et al., 2011; Construction Industry Institute (CII), 2008). As for the schedule performance, the empirical study of Hwang and Leong (2013) found that 16% of conventional construction projects were delayed while 32% of green construction projects were completed behind schedule. Thus, productivity improvement for green buildings is imperative, particularly for the sustainable development of green buildings in the long run. The construction industry needs to not only quickly adopt eco-friendly practices and materials that reduce its environmental impact but also take its own initiatives and find alternative ways to build green buildings in a more productive manner. In order to avoid fruitless efforts, the most important thing is to research the productivity performance and the causes affecting project productivity of green construction projects before making any improvement effort.

12.2 METHOD AND DATA COLLECTION

The questionnaire survey method and criticality index method were adopted in this chapter, and both methods have already been elaborated in Chapter 3. Prior to developing the questionnaire, a comprehensive literature review was conducted to identify the factors that may affect the productivity of green construction projects and to identify the solutions that may improve the productivity of green construction projects. As indicated in Tables 12.1 and 12.2, 26 productivity factors and 12 improvement solutions were identified from the literature review, which were then utilized to develop the questionnaire. In this chapter, the questionnaire survey was carried out to: (1) investigate and compare the productivity performance of green and conventional construction projects; (2) identify the significant factors affecting the productivity of green construction projects with comparison against traditional construction projects; and (3) propose feasible solutions to improve the productivity of green construction projects.

To ensure the correctness and accuracy of the questionnaire, presurvey interviews were conducted with three experts from the local green construction industry. The finalized questionnaire first explained the definitions of three

TABLE 12.1 Factors Affecting Construction Productivity

Factor Category	Code	List of Factors	References														
			1	*2*	*3*	*4*	*5*	*6*	*7*	*8*	*9*	*10*	*11*	*12*	*13*	*14*	*15*
Project factors	P1	Procurement method	√														
	P2	Construction method		√	√												
	P3	Timely payments				√											
	P4	Proportion of outsourced work					√										
	P5	Reworks						√	√								
	P6	Building design					√			√							
Manpower factors	MP1	Motivation of workers					√				√	√					√
	MP2	Workers' skill level										√				√	√
	MP3	Absenteeism						√	√				√				
	MP4	Labor turnover											√				
	MP5	Workers' experience										√				√	√
	MP6	Difficulty in recruitment of workers											√				

Continued

TABLE 12.1 Factors Affecting Construction Productivity—cont'd

Factor Category	Code	List of Factors	References														
			1	*2*	*3*	*4*	*5*	*6*	*7*	*8*	*9*	*10*	*11*	*12*	*13*	*14*	*15*
Management factors	MG1	Supervision of labor				√	√										
	MG2	Planning and sequencing of work									√	√					
	MG3	Competency of project manager				√	√										√
	MG4	Poor site layout				√			√								
	MG5	Inspection delay				√											
	MG6	Communication of information				√	√						√				
	MG7	Poor instructions				√			√								
Technical factors	T1	Materials availability		√				√	√								√
	T2	Tools and equipment		√				√	√								
	T3	Design changes						√	√								
	T4	Incomplete design				√											
	T5	Technology												√			
External factors	E1	Industry initiatives										√					
	E2	Weather			√								√		√		

References: 1. Dulaimi and Dalziel (1994); 2. Olomolaiye and Ogunlana (1989); 3. Thomas et al. (1999); 4. Makulsawatudom et al. (2004); 5. Jarkas and Bitar (2014); 6. Olomolaiye et al. (1987); 7. Kaming et al. (1997); 8. Fox et al. (2002); 9. Chan (2002); 10. Mojahed and Aghazadeh (2008); 11. Lim and Alum (1995); 12. Goodrum and Haas (2002); 13. Chan et al. (2013); 14. Han et al. (2008); 15. El-Gohary and Aziz (2014).

TABLE 12.2 Solutions to Improve the Productivity Performance of Green Construction Projects

Solution Category	Code	Solutions to Improve the Productivity of Green Construction Project	References
Solutions for project factors	SP1	To have building designs take into consideration standardized designs such as prefabricated components and materials which are readily and quickly available instead of specialized designs	Fox et al. (2002)
	SP2	To analyze the project requirements before the procurement phase so as to pick the right procurement method for the project	Mollaoglu-Korkmaz et al. (2013)
	SP3	Checking for mistakes in construction, construction method and quality control that would lead to defects so to avoid the possibility of reworks	Kaming et al. (1997)
Solutions for manpower factors	SMP1	Compensate workers with better wages so as to increase worker motivation	Mahamid (2013)
	SMP2	Increase incentives for companies to send their workers for skills improvement and retraining	Mahamid (2013)
	SMP3	Hiring of workers on a multiple projects or long term basis so as to present better job prospects and job security for workers, reducing labor turnover and increasing retention rate	Chan (2002)
Solutions for technical factors	ST1	Investing in new tools and equipment to improve the quality, maintenance and level of technology of tools for work	Kaming et al. (1997), Olomolaiye and Ogunlana (1989)

Continued

TABLE 12.2 Solutions to Improve the Productivity Performance of Green Construction Projects—cont'd

Solution Category	Code	Solutions to Improve the Productivity of Green Construction Project	References
	ST2	Clients making timely progress payment to contractors as it affects the contractors' ability to finance the work	Makulsawatudom et al. (2004)
	ST3	Check for mistakes and discrepancies in designs so to avoid the possibility of changes and corrections in design	Jarkas and Bitar (2014)
Solutions for management factors	SMG1	Development of plans such as a communication plan to improve communication of instructions and a work break down structure to improve coordination and sequencing	Hewage and Ruwanpura (2006), Robichaud and Anantatmula (2011)
	SMG2	Conducting good sourcing instead of cheap sourcing in the employment of foreman and supervisors	Lim and Alum (1995)
	SMG3	Consultants or superintendents to not delay the checking, reviewing and approving of design documents and necessary checks	Hewage and Ruwanpura (2009)

critical terms to the respondents, which is a crucial premise for the comparison between green and conventional construction projects. Productivity refers to the effective and efficient use of resources in achieving a desired output. Green buildings are buildings that have obtained the Singapore Green Mark certification, while conventional buildings are those that did not obtain the certification. The questionnaire then included the questions meant to profile the companies and respondents.

Afterward, the respondents were asked to indicate the number of green and conventional construction projects carried out by the company in the past 5 years. In order to investigate the severity of the impact of the productivity issues, the respondents were also asked to indicate the number of projects that experienced productivity issues and the number of projects delayed

TABLE 12.3 Five-Point Likert Scales for Likelihood and Impact

Likelihood			Impact	
L	*Linguistic Terms*	*Likelihood References (%)*	*I*	*Linguistic Terms*
1	Least likely	<20	1	Very low
2	Less likely	20–40	2	Low
3	Neutral	40–60	3	Medium
4	Likely	60–80	4	High
5	Most likely	>80	5	Very high

due to these issues. Moreover, after presenting the 26 factors affecting productivity in the questionnaire, the respondents were subsequently asked to assess the likelihood and impact of the factors affecting productivity in green and conventional construction projects, respectively, using the five-point scale elaborated in Table 12.3. At the end, the respondents were asked to rate the usefulness of the solutions using another five-point scale (1 = useless; 2 = somewhat useful; 3 = neutral; 4 = useful; 5 = very useful). In addition, postsurvey interviews were conducted with three industry experts who possess experience in both conventional and green construction projects. In postsurvey interviews, the experts were provided with the results obtained from the survey questionnaire. They all confirmed that the findings of the survey were reasonable and consistent with their expectations, which helped validate the findings from the survey. In order to gain an in-depth understanding of the findings, they were also requested to provide some possible explanations for the results.

The target population of the questionnaire survey consisted of all the professionals from the BCA-registered contractors and licensed builders, professionals from the firms under the Singapore Institute of Surveyors and Valuers (SISV) and project managers with Green Mark Certification, such as Green Mark Managers (GMM) and Green Mark Professionals (GMP). A total of 130 sets of survey questionnaires were randomly sent out to gather responses from developers, contractors and consultants. Finally, 32 complete sets were received, representing a response rate of 24.6%. Although the sample size was not large, statistical analysis could still be performed because the central limit theorem holds true when the sample size is no less than 30 according to the generally accepted rule (Hwang et al., 2015a; Ott and Longnecker, 2010; Zhao et al., 2013a).

The profiles of the respondents and companies are provided in Table 12.4. The respondents consisted of 16 project managers (PM), 13 quantity surveyors (QS) and 3 team members (TM) from 8 contractors, 12 consultants, and 12 developer firms.

TABLE 12.4 Profiles of Respondents and Companies

Characteristics		Frequency	Percentage	Years of Experience					
				Construction Industry			*Green Building Projects*		
				1–2	*3–4*	*≥5*	*1–2*	*3–4*	*≥5*
Respondent	PM	16	50.0%	2	1	13	6	2	8
	QS	13	40.6%	9	1	3	9	2	2
	TM	3	9.4%	0	0	3	2	0	1
	Subtotal	32	100%	11	2	19	17	4	11
	% by year			34.4%	6.3%	59.3%	53.1%	12.5%	34.4%
Company	Contractor	8	25.0%	0	0	8	0	4	4
	Consultant	12	37.5%	0	0	12	0	0	12
	Developer	12	37.5%	0	1	11	0	1	11
	Subtotal	32	100%	0	1	31	0	5	27
	% by year			0%	3.1%	96.9%	0%	15.6%	84.4%

In order to investigate the productivity performance, the respondents were asked to record the number of conventional and green construction projects which were performed over the past 5 years. The profiles of the projects are provided in Table 12.5.

From the perspective of project type, a total of 841 projects were recorded, comprising 540 new construction projects (64.21%) and 301 addition & alteration projects (35.79%). From the perspective of project size, the projects constructed ranged from S$1 million to S$51 million and above. The main bulk (54.22%) of the projects were in the lower tier of S$1M to S$20M, with the middle tier of S$21M to S$40M and higher tier of S$41M and above, having percentages of 21.05% and 24.73%, respectively. Among the projects, the majority were green building projects (61.12%).

12.3 ANALYSIS RESULTS AND DISCUSSIONS

12.3.1 Productivity Performance of Green Construction Projects

In order to compare the productivity performance, the projects were first analyzed on the basis of project type considering the frequency of occurrence (hereinafter referred to FO) of productivity issues by using a Chi-square test. The Chi-square test is first used to test whether FO is equally distributed between green and conventional construction projects under each project type. The null hypothesis H_0 for this test is that FO is statistically equally distributed between green and conventional construction projects under each project type. The Chi-square test is further used to test the independence between project type (NC and A&A) and building construction method (conventional and green construction). The null hypothesis H_0' for this test is that FO is statistically independent between project type and building construction method.

The test was conducted on the number of projects experiencing productivity issues and the number of projects delayed caused by these issues. The confidence level for both tests was 95% with a P-value of .05. If a P-value is less than .05, the null hypothesis is rejected. The analysis results of productivity performance by project type are summarized in Table 12.6.

Under principle 1, the overall percentage of green construction projects that had experienced productivity issues (42.8%) was much higher than that of conventional construction projects (22.0%). From the perspective of equal distribution, all P-values (0^*) under each project type were less than .05, which indicated that the FO of projects with productivity issues were statistically dependent on whether the project is green or conventional. From the perspective of independence, the H_0' was rejected because the P-value was zero, which indicated the dependence of FO of projects with productivity issues between project type and building construction method was statistically significant.

Green construction projects more frequently experienced productivity issues than conventional construction projects first because the two types of

TABLE 12.5 Profile of Projects

Characteristics of Projects		Traditional Projects		Green Projects		Total Frequency	Total Percentage
		Frequency	*Percentage*	*Frequency*	*Percentage*		
Type	NC	195	59.63%	345	67.12%	540	64.21%
	A&A	132	40.37%	169	32.88%	301	35.79%
Size (S$)	1M–10M	188	57.49%	107	20.82%	295	35.08%
	11M–20M	52	15.90%	109	21.21%	161	19.14%
	21M–30M	24	7.34%	78	15.18%	102	12.13%
	31M–40M	14	4.28%	61	11.87%	75	8.92%
	41M–50M	17	5.20%	74	14.40%	91	10.82%
	≥51M	32	9.79%	85	16.54%	117	13.91%
	Total	327	38.88%	514	61.12%	841	100.00%

Note: A&A, Addition & Alteration Project; *NC,* New Construction Project.

TABLE 12.6 Analysis Result of Productivity Performance by Project Type

	Building Construction Method			
Project Type	*No. of Conventional*	*No. of Green*	Total	*P*-Value (Chi-Square Test)
Principle 1: Project with productivity issues				
NC	65	163	228	*0**
A&A	7	57	64	*0**
Total	72	220	292	0
Principle 2: Project delayed caused by productivity issues				
NC	33	129	162	*0**
A&A	6	41	47	*0**
Total	39	170	209	.395
Total no. of projects	327	514	841	–
Overall statistic of productivity performance (%)				
Principle 1	22.0%	42.8%	34.7%	–
Principle 2	54.2%	77.3%	71.6%	–

projects have significant differences, such as design and contracting, as explained above. Furthermore, firms undertaking green construction projects may have less experience and knowledge in them because the green building concept has existed for a relatively short period of time, and this may cause a lack of productivity. Respondents in the postsurvey interview also provided a possible explanation that the professionals' unfamiliarity with green construction projects as compared to conventional construction projects resulted in them being more unproductive.

Under principle 2, among the projects that experienced productivity issues, the percentage of green construction projects delayed caused by productivity issues (77.3%) was also much higher than that of conventional construction projects (54.2%). From the perspective of equal distribution, all *P*-values (*0**) for each project type were less than .05, which indicated that the FO of projects delayed by productivity issues were statistically dependent on whether the project is green or conventional. The higher frequency of green construction projects experiencing delays caused by productivity issues is also supported by Hwang and Leong (2013). However, the $H_0{}'$ was not rejected because the *P*-value (.395) was greater than .05, which indicated that the dependence of

FO of projects delayed caused by productivity issues between project type and building construction method was not statistically significant. This, however, could be due to the low project count for A&A projects (only six projects).

Similarly, Chi-square test analysis was also conducted on the basis of project size. The Chi-square test is first used to test whether FO is equally distributed between the green and conventional construction projects under each project size. The Chi-square test is further used to test the independence between project size and building construction method. The confidence level for both tests was 95% with a P-value of .05. If a P-value is less than .05, the null hypothesis is rejected. The analysis results of productivity performance by project size are summarized in Table 12.7.

Under principle 1, except for projects between S$1M and S$10M, the equal distribution of FO of projects with productivity issues between green and conventional construction projects under each project size should be rejected because all P-values of all the other project sizes (*0**) were less than .05. Furthermore, the dependence of FO of projects with productivity issues between project size and building construction method was statistically significant because the P-value was zero. As for the difference in FO between conventional and green construction projects, projects more than S$51M had the greatest difference (35.3%) followed by projects between S$31M and S$40M (31.6%).

Under principle 2, the equal distribution of FO of projects delayed caused by productivity issues between green and conventional construction projects under each project size should be rejected because all P-values were less than .05. Furthermore, the dependence of FO of projects delayed caused by productivity issues between project size and building construction method was statistically significant because the P-value was zero. As for the difference in FO between conventional and green construction projects, projects more than S$51M had the greatest difference (66.7%) followed by projects between S$31M and S$40M (64.3%). For large projects, the higher difference could be due to the larger job scope which may result in more uncertainty leading to more productivity issues, such as reworks and design changes.

12.3.2 Factors Affecting Productivity Performance of Green Construction Projects

The following sections provide the results of likelihood, impact, and criticality index, respectively. Although these three results are mutually connected, analyses of these factors from different dimensions can further explore the causes of the productivity problem of green construction projects and help practitioners take different mitigation strategies. For example, for a factor with a high likelihood, a mitigation strategy should focus on reducing the likelihood that the factor will occur; while for a factor with a high impact, a mitigation strategy should focus on reducing the impact that the factor will have if it does occur and having contingency plans in place in case it does.

TABLE 12.7 Analysis Result of Productivity Performance by Project Size

Project Size (S$)	Building Construction Method		Total	*P*-Value (Chi-Square Test)	Difference in FO (*G* − *T*)
	No. of Conventional (T)	*No. of Green (G)*			
Principle 1: Project with productivity issues					
1M–10M	28	27	55	*1**	10.3%
11M–20M	28	60	88	*0**	1.2%
21M–30M	4	37	41	*0**	30.7%
31M–40M	2	28	30	*0**	31.6%
41M–50M	7	30	37	*0**	−0.7%
≥51M	3	38	41	*0**	35.3%
Total	72	220	292	0	20.8%
Principle 2: Project delayed caused by productivity issues					
1M–10M	9	24	33	*.01**	56.8%
11M–20M	24	50	74	*0**	−2.4%
21M–30M	4	22	26	*0**	−40.5%
31M–40M	0	18	18	*0**	64.3%
41M–50M	1	18	19	*0**	45.8%
≥51M	1	38	39	*0**	66.7%
Total	39	170	209	0	23.0%

Let us turn to the likelihoods of various factors by adopting methods, such as descriptive means, one sample *t*-test, paired *t*-test, and Spearman rank correlation, in order to compare the factors between green and conventional construction projects. First, the one sample *t*-test was used to test the mean scores of the likelihood of the factors against the mean value of three, which is the middle value of the scale. Furthermore, the paired *t*-test was used to test whether the means of the likelihood of the factors for the green construction projects were statistically different from those for the conventional projects. Finally, the Spearman rank correlation, which is a method of computing a correlation between the ranks of scores between two groups, was performed to test for any agreement on the ranking of factors between conventional and green construction projects. The test results of the likelihood of the factors are shown in Table 12.8.

TABLE 12.8 Test Results of the Likelihood of Factors

Factor Category	Code	List of Factors	Green (*G*)			Traditional (*T*)			Diff.	SD	*P* (p)
			Mean	*Rank*	P (*o*)	*Mean*	*Rank*	P (*o*)			
Project factors	P1	Procurement method	3.56	20	.00	3.47	18	.00	0.09	0.64	.41
	P2	Construction method	4.22	5	.00	4.19	2	.00	0.03	0.54	.75
	P3	Timely payments	2.97	26	.44	2.91	26	.29	0.06	0.56	.54
	P4	Proportion of outsourced work	3.53	21	.00	3.41	24	.01	0.12	0.94	.46
	P5	Reworks	3.72	17	.00	3.81	9	.00	−0.09	0.86	.54
	P6	Building design	4.13	8	.00	3.88	6	.00	0.25	0.72	.06
Manpower factors	MP1	Motivation of workers	3.41	24	.01	3.44	22	.00	−0.03	0.18	.33
	MP2	Workers' skill level	4.22	5	.00	4.03	3	.00	0.19	0.59	.08
	MP3	Absenteeism	3.41	25	.02	3.31	25	.04	0.10	0.30	.08
	MP4	Labor turnover	3.50	23	.00	3.44	22	.00	0.06	0.35	.33
	MP5	Workers' experience	4.44	1	.00	4.19	1	.00	0.25	0.67	.04
	MP6	Difficulty in recruitment of workers	3.97	11	.00	3.47	18	.00	0.50	0.92	.00

Management factors	MG1	Supervision of Labor	3.91	13	.00	3.59	16	.00	0.32	0.64	.01
	MG2	Planning and sequencing of work	4.19	7	.00	3.91	5	.00	0.28	0.58	.01
	MG3	Competency of project manager	4.28	2	.00	3.94	4	.00	0.34	0.65	.00
	MG4	Poor site layout	3.94	12	.00	3.84	8	.00	0.10	0.39	.18
	MG5	Inspection delay	3.66	19	.00	3.47	18	.00	0.19	1.06	.33
	MG6	Communication of information	3.88	14	.00	3.63	13	.00	0.25	0.67	.04
	MG7	Poor instructions	3.72	18	.00	3.66	12	.00	0.06	0.44	.42
Technical factors	T1	Materials availability	3.81	16	.00	3.56	17	.00	0.25	0.57	.02
	T2	Tools and equipment	3.84	15	.00	3.47	18	.00	0.37	0.75	.00
	T3	Design changes	4.25	4	.00	3.78	10	.00	0.47	0.67	.00
	T4	Incomplete design	4.06	9	.00	3.63	13	.00	0.43	0.76	.00
	T5	Technology	4.28	2	.00	3.69	11	.00	0.59	1.01	.00
External factors	E1	Industry initiatives	4.03	10	.00	3.88	6	.00	0.15	0.51	.10
	E2	Weather	3.53	22	.00	3.63	13	.00	−0.10	0.73	.48

Note: P (o), P-value in one sample t-test; P (p), P-value in paired t-test; Diff., difference in mean; SD, standard deviation.
The sample size of t-test is 32. The rank correlation between green and traditional is .86 with P-value of .00.

From the results of the one sample t-test, the mean scores of the likelihood of all factors except "Timely Payments" were statistically greater than the test value of three with respect to both conventional and green construction projects. This result indicated that these factors were significantly likely to affect the project productivity. For "Timely Payments," the likelihood of affecting project productivity was not statistically significant. This could be due to the Singapore Security of Payments Act (SOPA), introduced to ensure timely payments to contractors and provide secured cash flow for projects. The objectives of this act on improving the cash flow to contractors and the effectiveness of adjudication introduced under this act were addressed by Teo (2008).

From the results of mean, the top three factors most likely to affect the productivity of green building construction projects were "workers' experience," "competency of project manager," and "technology." "Workers' experience" was ranked first ($L=4.44$). First, the workers of the construction industry are heavily reliant on labor, which is more casual in nature (Debrah and Ofori, 1997; Jarkas and Bitar, 2014). As such, the experience of workers is most likely to affect the productivity. Furthermore, because green construction projects have only been around for the last 10 years, the workers would have limited experience. Hence, the likelihood of the productivity of green construction projects being affected by the workers' experience is high.

"Competency of project manager" was ranked second ($L=4.28$). According to Frank (2001), a project manager has a direct influence over 34%–47% of a project's success. From procurement to construction, green construction projects need to have more competent project managers as such projects are newer and more difficult to integrate. A project manager fulfills not only the traditional roles of project management but also must manage a project in the most efficient and effective manner with respect to green construction projects. A research study substantiated that the different knowledge areas of green construction affected the competency of project manager in comparison to conventional construction (Hwang and Ng, 2013). Moreover, the need for competent specialists increases with the increase in the number of green construction projects. In light of this, the BCA of Singapore introduced the Certified Green Mark Professional program to expand the industry's capability in the area of Sustainable Design and Development (BCA, 2016b).

"Technology" received the third position ($L=4.28$). Green construction specifications and methods differ from those for conventional construction. Green technologies and techniques are critical for such specifications to be drafted and implemented effectively (Lam et al., 2010). Although green technologies can offer plenty of sustainable solutions to help buildings and the environment, practitioners still need time to learn them and gain proficiency in these technologies. Furthermore, interviewees from postsurveys felt that green construction projects had lower productivity than conventional construction projects. They thought a possible explanation was the unfamiliarity with green technologies and technical difficulties during the construction process. These causes have been also indicated as two of the main challenges faced in green construction projects (Zhang et al., 2011; Tagaza and Wilson, 2004).

From the result of the Spearman rank correlation, it can be inferred that the factors likely to affect productivity in green construction projects have a strong positive correlation with the factors in conventional projects because the correlation value was .86 and the P-value was less than .05. However, from the results of the paired t-test, 11 factors were considered to be significant as the P-values of these factors were below .05. As such, the null hypothesis for the 11 factors should be rejected, indicating that the differences in likelihood between conventional and green construction projects for these factors were statistically significant. It is worth noting that five factors among the 11 were technical factors, representing that people think that green construction projects are more difficult in technology than conventional ones. From the results of the mean difference, "difficulty in recruitment of workers" (Diff. = 0.50), "supervision of labor" (Diff. = 0.32), "competency of project manager" (Diff. = 0.34), "tools & equipment" (Diff. = 0.37), "design changes" (Diff. = 0.47), "incomplete design" (Diff. = 0.43), and "technology" (Diff. = 0.59) were all factors having higher mean difference values (Diff. $\geq$0.30). In light of the above, factors with a high likelihood and factors with a high mean difference should draw the attention of practitioners.

The impacts of the various factors are also further analyzed by using the same methods mentioned above. The test results of the impact of the factors are shown in Table 12.9.

From the results of the one sample t-test, the mean scores of the impact of all factors except "absenteeism" were statistically greater than the test value of three with respect to both conventional and green construction projects. The results indicated that these factors had a significant impact on the project productivity. For "absenteeism," although the means for green ($I = 3.3$) and traditional ($I = 3.09$) were both greater than three, the P-values were also greater than .05, indicating that the impact of this factor was not statistically significant. The negation of the impact of "absenteeism" on productivity could be due to the abundance of cheap foreign labor in Singapore, which results in easy recruitment when there is absenteeism.

From the results of the mean, the top three factors having a high impact on the productivity of green construction projects were "workers' experience," "technology," and "design changes." "Workers' experience" was ranked first ($I = 4.41$). Due to the use of casual workers in green construction projects,

TABLE 12.9 Test Results of the Impact of Factors

Factor Category	Code	List of Factors	Green (*G*)			Traditional (*T*)			Diff.	SD	*P* (p)
			Mean	*Rank*	P *(o)*	*Mean*	*Rank*	P *(o)*			
Project factors	P1	Procurement method	3.81	15	.00	3.66	15	.00	0.15	0.45	.06
	P2	Construction method	4.16	7	.00	4.09	1	.00	0.07	0.44	.42
	P3	Timely payments	3.50	21	.00	3.53	19	.00	−0.03	0.18	.33
	P4	Proportion of outsourced work	3.56	20	.00	3.47	21	.00	0.09	0.78	.50
	P5	Reworks	4.03	12	.00	4.00	6	.00	0.03	0.40	.66
	P6	Building design	4.25	5	.00	4.06	2	.00	0.19	0.64	.11
Manpower factors	MP1	Motivation of workers	3.38	24	.02	3.44	22	.01	−0.06	0.35	.33
	MP2	Workers' skill level	4.22	6	.00	3.84	9	.00	0.38	0.71	.01
	MP3	Absenteeism	3.03	26	.43	3.09	26	.29	−0.06	0.56	.54
	MP4	Labor turnover	3.38	24	.01	3.25	25	.04	0.13	0.61	.26
	MP5	Workers' experience	4.41	1	.00	4.06	2	.00	0.35	0.65	.01
	MP6	Difficulty in recruitment of workers	3.75	17	.00	3.41	23	.00	0.34	0.60	.00

Management factors	MG1	Supervision of labor	3.72	19	.00	3.59	17	.00	0.13	0.61	.26
	MG2	Planning and sequencing of work	4.28	4	.00	4.03	4	.00	0.25	0.76	.07
	MG3	Competency of project manager	4.06	11	.00	4.00	6	.00	0.06	0.44	.42
	MG4	Poor site layout	3.75	17	.00	3.53	19	.00	0.22	0.49	.02
	MG5	Inspection delay	3.41	23	.00	3.28	24	.02	0.13	0.66	.29
	MG6	Communication of information	3.81	15	.00	3.75	14	.00	0.06	0.62	.57
	MG7	Poor instructions	3.84	14	.00	3.66	15	.00	0.18	0.54	.06
Technical factors	T1	Materials availability	4.13	8	.00	3.84	9	.00	0.29	0.52	.01
	T2	Tools and equipment	4.00	13	.00	3.78	13	.00	0.22	0.42	.01
	T3	Design changes	4.31	3	.00	4.03	4	.00	0.28	0.52	.01
	T4	Incomplete design	4.13	8	.00	3.84	9	.00	0.29	0.52	.01
	T5	Technology	4.38	2	.00	3.81	12	.00	0.57	0.76	.00
External factors	E1	Industry initiatives	4.09	10	.00	3.91	8	.00	0.18	0.69	.14
	E2	Weather	3.50	21	.00	3.56	18	.00	−0.06	0.56	.54

Note: The sample size of t-test is 32. The rank correlation between green and traditional is .90 with P-value of .00.

workers' inexperience can affect productivity through mistakes leading to reworks, changes, and delays in projects. Reworks could greatly affect a project's productivity from the aspects of cost and schedule (Hwang et al., 2009). Furthermore, the postsurvey interviewees disclosed that the market has a lack of workers with experience in green construction projects and with a green certificate. Although the BCA introduced the Certified Green Mark Professional program, some workers still hesitate to join the program while other workers have not finished the program.

"Technology" received the second position ($I=4.38$). Energy-efficient technologies could not only cut down on waste on site but also decrease the effort expended by all parties in the construction process, which is an essential productivity improvement. The impact of "technology" on productivity was also supported by Goodrum and Haas (2004), stating that improvements in technology in the US construction industry had increased productivity over the course of 25 years.

"Design changes," which are prevalent in the construction industry, was ranked third ($I=4.31$). In order to quickly finish a project, the time spent on the planning stage is usually not adequate. Taking Taiwan highway construction as an example, insufficient time and resources spent during the planning stage led to increased frequency of design changes (Wu et al., 2005). While design should be ironed out in green construction projects earlier on, failure to do so can lead to designs with little integration or to design changes during construction. The high number and various types of design changes can invariably result in a drop off in the efficiency in construction (Thomas and Napolitan, 1995).

From the result of the Spearman rank correlation, it can be inferred that the factors that have an impact on the productivity in green construction projects have a strong positive correlation with the factors in conventional projects because the correlation value was .90 and the P-value less than .05. However, from the results of the paired t-test, nine factors were considered to be significant as the P-values of these factors were below .05. As such, the null hypothesis for the nine factors should be rejected, indicating that the differences in impact on productivity between conventional and green construction projects for these factors were statistically significant. It is worth noting that the majority of the nine factors were from manpower factors (three) and technical factors (five). From the results of the mean difference, "workers' skill level" ($\text{Diff.}=0.38$), "workers' experience" ($\text{Diff.}=0.35$), "difficulty in recruitment of workers" ($\text{Diff.}=0.34$) and "technology" ($\text{Diff.}=0.57$), were factors having higher mean difference values ($\text{Diff.} \geq 0.30$). In light of the above, factors with a high impact and factors with a high mean difference should draw the attention of practitioners.

Using Eqs. (3.1), (3.2), the criticality index (CI) of each factor on the project productivity for conventional and green construction projects was calculated.

The criticalities of the various factors were analyzed by using the same methods mentioned above. One difference is that the test value of the one-sample *t*-test was nine. The test results of the criticality are shown in Table 12.10.

From the results of the one sample *t*-test, all factors were significantly critical to affecting the productivity of both conventional and green construction projects because all mean values were greater than nine and all *P*-values were less than .05. From the results of the mean, the top five factors (CI > 18) critical to affecting the productivity of green construction projects were "workers' experience," "technology," "design changes," "workers' skill level," and "planning & sequencing of work." The five factors are discussed below.

"Workers' experience" was ranked first (CI = 19.78). The highest likelihood and impact of this factor contributed to the top ranking. The possible reasons for the high likelihood and impact were discussed above. Moreover, the criticality of "workers' experience" was also demonstrated in Mojahed and Aghazadeh (2008) and El-Gohary and Aziz (2014) in which one of the top five factors critical to productivity was workers' experience. The view that "workers' experience" is critical to the productivity of green construction projects may also be because such projects are relatively new and experience is hard to come by.

"Technology" received the second position (CI = 19.13). The reasons for the high likelihood and impact of "technology" were discussed above. Moreover, different forms of technology employed would differ depending on the type of work done and level of technology employed by contractors in completing a task. Because the design and construction practices of green construction are more complex, due to a high level of integration and new green materials, new technologies have to be adopted and are critical for the productivity of green construction projects.

"Design changes" was ranked third (CI = 18.72). Insufficient knowledge, insufficient technical expertise, and unfamiliarity with the products, materials, systems or designs actually cause certain obstacles for developers, clients and contractors (Eisenberg et al., 2002), which may easily lead to design changes. Moreover, the highly-integrated nature of green technologies and their impact on other construction elements at the planning and design stage, are very important. A failure to take integration into account may result in construction conflicts and design changes. Design changes were identified as a cause for rework during the design or construction stages, both of which would cause a decrease in productivity (Kaming et al., 1997; Olomolaiye et al., 1987). In light of this, Building Information Modeling (BIM) was recommended for use in green construction projects (Wu and Issa, 2014). The criticality of "design changes" was also supported by Kaming et al. (1997) and Olomolaiye et al. (1987), with both indicating the importance of design changes.

TABLE 12.10 Test Results of the Criticality of factors

Factor Category	Code	List of Factors	Green (*G*)			Traditional (*T*)			Diff.	SD	*P* (p)
			Mean	*Rank*	P *(o)*	*Mean*	*Rank*	P *(o)*			
Project factors	P1	Procurement method	13.97	19	.00	12.94	19	.00	1.03	2.87	.05
	P2	Construction method	17.91	6	.00	17.53	1	.00	0.38	2.92	.47
	P3	Timely payments	11.00	25	.03	10.78	26	.00	0.22	1.95	.53
	P4	Proportion of outsourced work	13.06	20	.00	12.34	20	.00	0.72	5.91	.50
	P5	Reworks	15.44	13	.00	15.63	7	.00	−0.19	4.51	.82
	P6	Building design	17.75	7	.00	16.09	4	.00	1.66	4.53	.05
Manpower factors	MP1	Motivation of workers	11.94	24	.00	12.28	21	.00	−0.34	1.94	.33
	MP2	Workers' skill level	18.31	4	.00	16.03	5	.00	2.28	4.17	.00
	MP3	Absenteeism	10.91	26	.04	10.81	25	.00	0.10	1.96	.79
	MP4	Labor turnover	12.22	23	.00	11.56	24	.00	0.66	3.02	.23
	MP5	Workers' experience	19.78	1	.00	17.34	2	.00	2.44	4.46	.00
	MP6	Difficulty in recruitment of workers	15.28	14	.00	12.09	22	.00	3.19	5.58	.00

Management factors	MG1	Supervision of labor	15.00	16	.00	13.22	18	.00	1.78	4.14	.02
	MG2	Planning and sequencing of work	18.19	5	.00	16.19	3	.00	2.00	4.13	.01
	MG3	Competency of project manager	17.69	8	.00	16.00	6	.00	1.69	3.61	.01
	MG4	Poor site layout	15.13	15	.00	14.00	13	.00	1.13	3.11	.05
	MG5	Inspection delay	12.91	22	.00	11.78	23	.00	1.13	6.14	.31
	MG6	Communication of information	15.00	16	.00	13.84	15	.00	1.16	3.92	.11
	MG7	Poor instructions	14.94	18	.00	14.00	13	.00	0.94	3.69	.16
Technical factors	T1	Materials availability	16.50	11	.00	14.38	11	.00	2.12	3.83	.00
	T2	Tools and equipment	15.66	12	.00	13.28	17	.00	2.38	3.92	.00
	T3	Design changes	18.72	3	.00	15.59	8	.00	3.13	4.56	.00
	T4	Incomplete design	17.28	9	.00	14.31	12	.00	2.97	4.82	.00
	T5	Technology	19.13	2	.00	14.44	10	.00	4.69	6.05	.00
External factors	E1	Industry initiatives	16.94	10	.00	15.53	9	.00	1.41	4.46	.08
	E2	Weather	13.00	21	.00	13.47	16	.00	−0.47	4.23	.54

Note: The sample size of t-test is 32. The overall rank correlation between green and traditional is .89 with P-value of .00.

"Workers' skill level" was ranked fourth (CI = 18.31). Burleson et al. (1998) indicated that the multiskilling of workers could improve overall project productivity through the potential cost saving in labor, reduction in workforce, and an average increase in employment duration. However, labor-related issues cannot be ignored currently. Lack of technical skill regarding green technologies and techniques, workers' lack of awareness of the correct methods and procedures, and the resistance to change from conventional practices, were indicated as three labor-related challenges in green construction projects (Hwang and Ng, 2013). Based on the acknowledgment of the criticality of "workers' skill level," productivity movements have been launched by the construction industry in Singapore targeting improvement in skills in order to improve productivity (BCA, 2011). Under these movements, more training on green construction should be provided to the workers. The postsurvey interviewees also emphasized the lack of capability of labor in Singapore. The cheap foreign labor can handle just small tasks and are always waiting for instructions from supervisors, resulting in low productivity.

"Planning & sequencing of work" was the fifth most critical factor (CI = 18.19). The lengthy planning and approval process for new green technologies and recycled material was pointed out as one of the main challenges faced in green construction projects (Zhang et al., 2011; Tagaza and Wilson, 2004). From this perspective, making a good plan and sequencing of work by higher management can lead to improvements in productivity, which is also supported by Chan (2002) and Mojahed and Aghazadeh (2008). Moreover, Liu et al. (2011) also proved that productivity does improve when workflow is made more predictable. Furthermore, implementing green construction practices onsite needs more time because green technologies require complicated techniques and construction processes (Hwang and Tan, 2012; Zhang et al., 2011). Therefore, more effort is needed to integrate green construction practices into the planning and sequencing of work than for conventional projects. Otherwise, more resources would have to be allocated to keep a project on schedule, which would cause a low project productivity.

The overall rank correlation was .89 with the P-value less than .05. This result indicates that the factors affecting the productivity in green construction projects have a strong positive correlation with the factors in conventional projects. However, from the results of the paired t-test, there were still 11 factors (almost half) considered to be significant as the P-values of these factors were below .05. As such, the null hypothesis for the 11 factors should be rejected, indicating that the differences in criticality between conventional and green construction projects for these factors were statistically significant.

Overall, the mean values for green construction projects generally tend to be higher than those for conventional projects in Tables 12.8–12.10. This means that most of the factors investigated have higher likelihoods and impacts for green construction projects than for conventional projects. It also possibly represents the respondents' perception about green construction projects. Factors with a high difference in criticality, such as "workers' skill level" (Diff. = 2.28), "workers' experience" (Diff. = 2.44), "difficulty in recruitment of workers" (Diff. = 3.19), "planning & sequencing of work" (Diff. = 2.0) and all technical factors (Diff. ≥2.12), are actually due to the perception that green construction projects are more difficult than conventional ones.

Due to the quick rise of green construction, there is an increasing need for green skilled workers. The different perceptions of "workers' skill level," "workers' experience," and "difficulty in recruitment of workers" could be due to the shortage of skilled workers across the green construction industry. Coupled with other factors, and the increase in green construction projects, it has led to a shortage of green construction workers, with 86% of architectural and engineering firms and 90% of general construction firms facing difficulties in recruitment (McGraw Hill Construction, 2012). The postsurvey interviewees also concurred with this viewpoint in the case of Singapore. As for technical factors, Lam et al. (2010) identified that green technologies, techniques and materials were the most important and difficult factors for the implementation of green specifications in construction. Green materials usually require special orders and manufacturing, greatly affecting the construction productivity (Kibert, 2008). However, the impact of technology on conventional project productivity was not emphasized in previous studies. In light of the above, it can be inferred that "technology" has a very different impact on the productivity of conventional and green construction projects. The designs for green constructions are more integrated and must consider many factors to ensure environmental friendliness (BCA, 2010a). Interviewees from the postsurvey agreed that the design had a greater emphasis in green construction than in conventional, resulting in the difference in criticality on productivity. In addition, green specifications and technologies have not been widely grasped by the majority of practitioners or contractors. Similarly, postsurvey interviewees stated that the difference between the tools used in green construction and conventional construction would result in a difference in productivity. In light of the above, factors with a high criticality and factors with a large mean difference should draw practitioners' attention. Meanwhile, these factors should be first considered when practitioners make specific adjustments to conventional project management practices and processes in green construction.

12.3.3 Solutions for Productivity Improvement of Green Construction Projects

The respondents were also asked to rate the usefulness of the 12 solutions to improve green construction productivity. The ranking solutions based on the frequency of usefulness and the ranking solutions based on the frequency of the top five selected recommendations are shown in Table 12.11.

The two types of analysis for the ranked solutions were used to identify similar solutions between objective and subjective ranked solutions. "To have

TABLE 12.11 Ranking of Solutions for Improving Green Construction Projects' Productivity

		Usefulness			Top 5	
Code	**Solutions Category**	*Mean*	*Rank*	*P-Value*	*Per.*	*Rank*
SP1	To have building designs take into consideration standardized designs such as prefabricated components and materials which are readily and quickly available instead of specialized designs	4.31	1	.00	43.8%	1
SP2	To analyze the project requirements before the procurement phase so as to pick the right procurement method for the project	4.00	5	.00	12.5%	9
SP3	Checking for mistakes in construction, construction method and quality control that would lead to defects so to avoid the possibility of reworks	4.00	5	.00	25.0%	4
SMP1	Compensate workers with better wages so as to increase worker motivation	3.34	12	.09	6.3%	10
SMP2	Increase incentives for companies to send their workers for skills improvement and retraining	4.00	5	.00	28.1%	3
SMP3	Hiring of workers on a multiple projects or long term basis so as to present better job prospects and job security for workers, reducing labor turnover and increasing retention rate	3.66	10	.00	3.1%	12

TABLE 12.11 Ranking of Solutions for Improving Green Construction Projects' Productivity—cont'd

		Usefulness			Top 5	
Code	Solutions Category	*Mean*	*Rank*	P-*Value*	*Per.*	*Rank*
ST1	Investing in new tools and equipment to improve the quality, maintenance and level of technology of tools for work	4.28	2	.00	34.4%	2
ST2	Clients making timely progress payment to contractors as it affects the contractors' ability to finance the work	3.56	11	.01	6.3%	10
ST3	Check for mistakes and discrepancies in designs so to avoid the possibility of changes and corrections in design	4.03	4	.00	25.0%	4
SMG1	Development of a plans such as a communication plan to improve communication of instructions and a work break down structure to improve coordination and sequencing	3.69	9	.00	21.9%	7
SMG2	Conducting good sourcing instead of cheap sourcing in the employment of foreman and supervisors	4.00	5	.00	25.0%	4
SMG3	Consultants or superintendents to not delay the checking, reviewing and approving of design documents and necessary checks	4.06	3	.00	21.9%	7

building designs take into consideration standardized designs such as prefabricated components and materials which are readily and quickly available instead of specialized designs" was ranked as the top one useful solution from both types of analysis. Designs should strive for ease of construction and buildability. Incorporating prefabricated components into the design could improve productivity. Both Low (2001) and Fox et al. (2002) showed that using designs with a high buildability score or producer-led standardized designs instead of customer-led specialized designs would result in improvements in productivity. This is due to the modules for materials and designs of components being quickly and readily available, allowing components to be quickly built and assembled, thus improving productivity. Interviewees for the postsurvey noted that such prefabrication of components may be easier to assemble on site and hence improve productivity.

In Singapore's construction productivity roadmap development in 2015, the implementation of prefabricated prefinished volumetric construction (PPVC) and prefabricated bathroom units (PBU) were emphasized. The Singapore government is encouraging the local construction industry to use more PPVC and PUB through public projects and state land scales (Tan, 2015). These kinds of policies and strategies will help in the improvement of productivity in green construction projects. However, there is a premise that the constructability level of the designs for green construction project should first be improved. This can be accomplished by increasing the designers' awareness of the significant impact of this concept on the productivity of the green construction process. This may, additionally, be augmented by encouraging procurement methods that allow the involvement of contractors during the design stage of projects, such as early contractor involvement (ECI) (Jarkas and Bitar, 2014).

"Investing in new tools and equipment to improve the quality, maintenance and level of technology of tools for work" was ranked second from both types of analysis. Investing in new tools and equipment, as well as maintenance, is crucial to improving productivity. This would give the workers the right tools needed for the job and increase the availability of tools by having them well maintained. This is also further substantiated by the fact that an improvement in technology would help to improve productivity overall (Goodrum and Haas, 2004). Furthermore, technology is seen as one of the top five factors critical to productivity and, as such, providing the right tools for the construction of green construction projects should be considered.

"Check for mistakes and discrepancies in designs so to avoid the possibility of changes and corrections in design" was the third similar and useful solution. Design changes have plagued the construction industry and it is a critical factor for green construction. One way to reduce such design changes is through the checking of mistakes and discrepancies in designs and design documents (Assaf and Al-Hejji, 2006). Kaming et al. (1997) expressed the view that detailed designs were crucial in preventing reworks. From this perspective, Building Information Modeling (BIM) software, which is well-known for the function of conflict checking, is recommended for use in green construction projects to improve the productivity (Wu and Issa, 2014).

"Increase incentives for companies to send their workers for skills improvement and retraining" received the fourth position. Workers' skills and experience were rated in the top five factors critical to productivity. While experience can only be gained through projects, skills can be learned and gained through training programs. By increasing the learning rate of workers, a more productive workforce can be quickly achieved. However, due to Singapore's labor being casual in nature, it is understandable why contractors are not willing to send them for training as they may leave at any time. Therefore, in order to increase the uptake of programs already present, incentives can be provided, such as cash back to contractors, so as to improve the skill of their workers and hence productivity. An incentive could be similar to the one suggested by Mahamid (2013), whereby the government itself provides the training programs at a lower cost so as to entice contractors to take the programs.

"Checking for mistakes in construction, construction method and quality control that would lead to defects so to avoid the possibility of reworks" was ranked the fifth useful solutions. Mistakes in construction are quite common. One reason for the occurrence of such mistakes is that the construction work performed is dissimilar for every project. Furthermore, the workers compound the problem due to poor coordination when handing over to the next section (Riemer, 1976). As such, checking of mistakes, such as integration plans in construction methods, before handing over to the next specialist, can ensure that the work is properly done to the correct standards thus decreasing the amount of rework.

12.4 SUMMARY

Through a comprehensive literature review and a questionnaire survey, this chapter investigated the difference in productivity performance between green and conventional construction projects. The significant factors affecting the productivity of green construction projects along with comparisons against conventional construction projects, and useful solutions to improve green construction productivity were also investigated.

The findings of this chapter first demonstrated that green construction projects had experienced productivity issues and delays caused by these issues more frequently than conventional construction projects. The results showed that only 22.0% of conventional construction projects had encountered productivity issues while green construction projects almost doubled that figure to 42.8%. Moreover, the percentages of projects delayed caused by productivity issues were 54.2% and 77.3% for conventional and green construction projects, respectively. This chapter further identified the top five factors affecting the productivity of green construction projects on which the industry should focus, and these were "workers' experience," "design changes," "technology," "workers' skill level," and "planning and sequencing of work." In addition, after comparing the significance indices of all factors between green and conventional construction projects, 11 factors were identified as having statistically significant difference, with "technology" having the largest difference. In addition, the results of this chapter indicated that all the solutions for improving the productivity of green construction projects were significantly useful but also showed the top three useful solutions to be "standardized design," "investment in new tools and equipment," and "consultants' or superintendents' on time checking, receiving and approving of design documents."

Chapter 13

Risk Assessment and Mitigation Measures for Green Construction Projects

13.1 INTRODUCTION

Construction projects are inevitably plagued with complex and diverse risks (Deng et al., 2014b; Hlaing et al., 2008; Zhao et al., 2014b). Similar to conventional construction projects, green construction projects face various risks (Ashuri and Durmus-Pedini, 2010). However, these may be different risks due to the use of new materials, technologies, and design approaches (Odom et al., 2008), and sustainable construction practices and achievement of third party green certification (Greenwald, 2012). If the risks associated with green projects are not appropriately managed, they will continue to become barriers to the green construction movement (Ashuri and Durmus-Pedini, 2010). Thus, risk management is necessary to ensure the success of green construction. The objectives of this chapter are to (1) identify the risk factors and groups associated with green construction projects in Singapore; (2) develop a risk assessment model using the fuzzy synthetic evaluation approach; and (3) assess the risk factors and groups using this model.

13.2 METHOD AND DATA PRESENTATION

As risk assessment is complex and ambiguous, qualitative linguistic terms are unavoidable (Wang et al., 2004). In addition, the perceptions on likelihood and impact of risk factors by respondents are typically subjective and uncertain (Shan et al., 2015a). The fuzzy set theory can deal with the problems relating to ambiguous, subjective and imprecise judgments (Zhao et al., 2013c; Pedrycz et al., 2011). The fuzzy set theory also allows mathematical operators to be applied to the fuzzy domain (Xia et al., 2011; Ma and Kremer, 2015), and can quantify the linguistic facet of available data and preferences for individual or group decision-making (Zimmermann, 2001; Zhao et al., 2014a). Thus, the fuzzy set theory is considered appropriate for risk assessment. As an application of the fuzzy set theory, fuzzy synthetic evaluation (FSE) aims to provide a synthetic evaluation of an object relative to an objective in a fuzzy decision

Performance and Improvement of Green Construction Projects. https://doi.org/10.1016/B978-0-12-815483-0.00013-2

environment with multiple criteria (Mu et al., 2014). FSE has been adopted in several risk management studies. For instance, Mu et al. (2014) applied this method to assess the risk management capability of contractors in subway projects in Mainland China; Xu et al. (2010a) developed a fuzzy synthetic risk allocation model for public-private partnership (PPP) projects in China; and Liu et al. (2013) used this method to analyze risks in ultra-deep scientific drilling projects. The advantage of FSE lies in its ability to deal with complicated evaluations with multiple attributes and levels (Xu et al., 2010a; Mu et al., 2014).

Risk management is necessary to ensure successful delivery of green projects as well as the achievement of key targets set by the government. A large number of studies have attempted to identify many risk factors in various construction projects. This chapter will summarize 28 risk factors that are common to all kinds of construction projects and those closely associated with green construction. The risk factors have also been categorized into 11 groups, as indicated in Table 13.1.

Based on the 28 risk factors identified from literature review, a questionnaire was developed to collect data. To ensure the quality of the questionnaire, presurvey interviews were conducted with three experienced experts from the local green construction industry to vet the questionnaire. All these experts have at least 10 years of experience on green construction projects, and their opinions can provide valuable insights into the risks in green construction projects. The finalized questionnaire contained two sections. The first section includes questions meant to profile the respondents and their firms. The second section requests the respondents to assess the likelihood of occurrence (LO) and the magnitude of impact (MI) of 28 risk factors using a five-point rating scale (1 = very low; 2 = low; 3 = medium; 4 = high; and 5 = very high).

The targeted population of the questionnaire was BCA-registered companies. According to the BCA (2015a), the registry grade represents different tendering limits for different contractors. Contractors with the registry grade of A1 have unlimited tendering limits while A2 indicates its limit at S$90 million. Grades B1 and B2 have tender limits of S$42 million and S$14 million, respectively. The target audience comprised the project managers experienced in green building from the companies with BCA financial grades A1 to B2. A total of 277 questionnaires were sent out to all the general building contractors registered with financial grades A1 to B2. The selection was restricted to financial grades A1 to B2 because their higher financial capabilities and resources allow them to undertake green projects. Moreover, a cross check on the green buildings constructed in Singapore revealed that most of these green projects were undertaken by larger companies classified under these categories.

A total of 31 completed responses were received, yielding a response rate of 11.2%, which was low compared with the norm of 20%–30% of most questionnaire surveys in the construction industry (Akintoye, 2000). The low response rate could result from the busy schedules of project managers that prevented

TABLE 13.1 Risk Factors in Green Construction Projects Identified From Previous Studies

Group	Code	Risk Factor	References									
			1	2	3	4	5	6	7	8	9	10
Macro-economic risk	R01	Increasing inflation rate	√	√			√	√	√	√	√	√
	R02	Currency exchange rate fluctuation	√	√	√	√	√		√	√		
	R03	Increasing tax rate		√		√			√			
Contract problems	R04	Unclear contract conditions for dispute resolution					√		√			√
	R05	Unclear contract conditions for claims and litigations	√									
Client-related risk	R06	Unclear requirements of clients	√			√						
	R07	Intervention of clients	√		√		√					
	R08	Delayed payments from clients			√		√			√		
Design problems	R09	Unclear detailed design or specifications	√	√		√	√					
	R10	Poor design	√			√	√		√		√	√
	R11	Variations in design	√	√		√	√	√		√	√	

Continued

TABLE 13.1 Risk Factors in Green Construction Projects Identified From Previous Studies—cont'd

Group	Code	Risk Factor	References									
			1	2	3	4	5	6	7	*8*	9	*10*
Safety risk	R12	Strict safety and health regulations	√	√	√	√						
	R13	Construction accidents	√		√			√	√		√	√
Procedure complexity	R14	Complex planning approval and permit procedures	√	√	√	√				√	√	
	R15	Delay in issuance of documents	√				√		√			
Technical problems	R16	Technical complexity	√			√	√		√	√		
	R17	Use of new construction method and technology		√	√		√	√	√			
Human resource risk	R18	Constraints on laborer employment	√	√	√	√						
	R19	Lack of management staff	√			√	√	√			√	
	R20	Poorly trained laborers	√		√			√			√	
Material and equipment problems	R21	Material and equipment problems R21 Material quality problems	√				√					
	R22	Availability of equipment and materials	√				√					
	R23	Default supply of materials, equipment and plants	√	√	√	√	√		√			√

Project team risk	R24	Project teams without relevant knowledge	√								√	
	R25	Inefficient communication and coordination	√					√			√	
	R26	Unfavorable sub-contractors	√		√		√				√	√
Cost overrun risk	R27	Inaccurate cost estimation	√					√				
	R28	Labor and materials price fluctuations	√		√							

Reference: 1. Zhi (1995); 2. Baker et al. (1999); 3. Hlaing et al. (2008); 4. Eybpoosh et al. (2011); 5. El-Sayegh (2008); 6. Lu and Yan (2013); 7. Panthi et al. (2009); 8. Al-Bahar and Crandall (1990); 9. Zou et al. (2007); 10. Kartam and Kartam (2001).

them from completing the surveys on time, and the confidentiality and sensitivity of information that companies were reluctant to divulge. Despite the small sample size, statistical analysis can still be applied because the central limit theorem holds true with a sample size >30 (Ott and Longnecker, 2001; Hwang et al., 2015c). Indeed, there has been no criterion on the sample size of FSE. For example, in the previous studies using FSE, the sample sizes were eight in Liu et al. (2013), six in Onkal-Engin et al. (2004), and 58 in Mu et al. (2014).

The profile of the companies and respondents is shown in Table 13.2. A total of 74.2% of the respondents were from A1 contractors and 77.4% had over 10 years of experience in the construction industry and more than 3 years of

TABLE 13.2 Profile of Companies, Respondents, and Projects

Characteristics			*N*	%
Company (*N*=31)	Registry grades[a]	A1	23	74.2%
		A2	5	16.1%
		B1	2	6.5%
		B2	1	3.2%
Respondent (*N*=31)	Experience in construction industry	<5 years	2	6.5%
		5–10 years	5	16.1%
		>10 years	24	77.4%
		Mean	*15.0*	–
	Experience in green building	<3 years	7	22.6%
		3–4 years	11	35.5%
		5–10 years	13	41.9%
		Mean	*6.7*	–
Project (*N*=114)	Type	Commercial	45	39.5%
		Residential	34	29.8%
		Education	23	20.2%
		Industry	8	7.0%
		Other	4	3.5%
	Nature	New construction	95	83.3%
		Addition and alteration	19	16.7%

[a]*BCA grading system: A1-unlimited tendering limit; A2-up to S$90 million; B1-up to S$42 million; B2-up to S$14 million.*

experience in green construction projects. Among the 114 green projects managed by the respondents, 69.3% were either commercial or residential projects, and 83.3% were new construction.

13.3 ANALYSIS RESULTS AND DISCUSSIONS

13.3.1 Risk Assessment Model Development and Risk Assessment Results

There are three levels of risks to be measured in green construction projects: risk factors (level 1), risk groups (level 2), and the overall risk (level 3). Therefore, it is appropriate to adopt FSE to develop the risk assessment model.

As a multicriteria evaluation model, the proposed risk assessment model requires three basic elements (Xu et al., 2010b; Mu et al., 2014):

(1) A set of basic criteria/factors $C=\{c_1,c_2,\ldots c_m\}$ where m is the number of criteria/factors;
(2) A set of alternatives $E=\{e_1,e_2,\ldots e_n\}$, where n is the number of alternatives; and
(3) An evaluation matrix $R=(r_{ij})_{m\times n}$, where r_{ij} denotes the degree to which the alternative e_j satisfies the criterion c_i.

The development of the proposed risk assessment model involves three steps:

Step 1: Calculating LO, MI, RC of risk factors
Step 2: Calculating LO, MI, RC of risk groups
Step 3: Calculating LO, MI, RC of overall risks

Step 1: Calculate LO, MI, and RC of Risk Factors (Level 1)

The LO and MI of each risk factor were collected in the questionnaire survey with a five-point scale. Thus, in the set of alternatives E, for both LO and MI, e_1 = very low; e_2 = low; e_3 = medium; e_4 = high; and e_5 = very high.

In the evaluation matrix, r_{ij} denotes the degree to which the alternative e_j satisfies the risk factor i. For example, the results on the LO of risk factor "Increasing inflation rate" (R01) indicate that 23% of the respondents rated the LO as very low, 26% as low, 35% as medium, 16% as high and 0% as very high. The membership function of the LO is given by Eq. (13.1):

$$\frac{0.23}{\text{very low}}+\frac{0.26}{\text{low}}+\frac{0.35}{\text{medium}}+\frac{0.16}{\text{high}}+\frac{0.00}{\text{very high}} =\frac{0.23}{1}+\frac{0.26}{2}+\frac{0.35}{3}+\frac{0.16}{4}+\frac{0.00}{5} \tag{13.1}$$

It can also be written as a matrix in Eq. (13.2):

$$\left(R_i^{\text{LO}}\right)_{1\times 5}=\left(r_{i1}^{\text{LO}},r_{i2}^{\text{LO}},r_{i3}^{\text{LO}},r_{i4}^{\text{LO}},r_{i5}^{\text{LO}}\right) \tag{13.2}$$

The LO and MI of risk factor i can be calculated using Eqs. (13.3), (13.4), respectively:

$$\mathrm{LO}_i = \sum_{j=1}^{5} \left(s_j \times r_{ij}^{\mathrm{LO}} \right) \tag{13.3}$$

$$\mathrm{MI}_i = \sum_{j=1}^{5} \left(s_j \times r_{ij}^{\mathrm{MI}} \right) \tag{13.4}$$

where s_j is the rating given to factor i, namely $s_j = 1, 2, 3, 4, 5$. In risk management studies, risk criticalities (RC) is usually adopted to measure how critical a risk factor is, and it is considered to be the product of LO and MI. To keep the scale of RC consistent with LO and MI, RC of risk factor i is calculated using Eq. (13.5):

$$\mathrm{RC}_i = \sqrt{\mathrm{LO}_i \times \mathrm{MI}_i} \tag{13.5}$$

Step 2: Calculate LO, MI, and RC of Risk Groups (Level 2)

To calculate the LO and MI of each risk group (level 2), the weight of each risk factor (level 1) within each risk group, $W = \{w_1, w_2, \ldots w_k\}$, should be determined. Here, k is the number of the risk factors within a risk group. The weights assigned to the LO and MI of risk factor i can be calculated by Eqs. (13.6), (13.7), respectively:

$$w_i^{\mathrm{LO}} = \mathrm{LO}_i \Big/ \sum_{i=1}^{k} \mathrm{LO}_i \tag{13.6}$$

$$w_i^{\mathrm{MI}} = \mathrm{MI}_i \Big/ \sum_{i=1}^{k} \mathrm{MI}_i \tag{13.7}$$

In FSE, the results of the evaluation are obtained by calculating the fuzzy composition of the weight vector W and the evaluation matrix R, namely $D = W \times R$. Thus, the LO and MI membership functions of risk group t can be calculated using Eqs. (13.8)–(13.11), respectively:

$$d_{tj}^{\mathrm{LO}} = \sum_{i=1}^{k} w_i^{\mathrm{LO}} \times r_{ij}^{\mathrm{LO}} \tag{13.8}$$

$$\left(D_t^{\mathrm{LO}}\right)_{1\times 5} = \left(W_i^{\mathrm{LO}}\right)_{1\times k} \times \left(R_i^{\mathrm{LO}}\right)_{k\times 5} = \left(d_{t1}^{\mathrm{LO}}, d_{t2}^{\mathrm{LO}}, d_{t3}^{\mathrm{LO}}, d_{t4}^{\mathrm{LO}}, d_{t5}^{\mathrm{LO}}\right) \tag{13.9}$$

$$d_{tj}^{\mathrm{MI}} = \sum_{i=1}^{k} w_i^{\mathrm{MI}} \times r_{ij}^{\mathrm{MI}} \tag{13.10}$$

$$\left(D_t^{\mathrm{MI}}\right)_{1\times 5} = \left(W_i^{\mathrm{MI}}\right)_{1\times k} \times \left(R_i^{\mathrm{MI}}\right)_{k\times 5} = \left(d_{t1}^{\mathrm{MI}}, d_{t2}^{\mathrm{MI}}, d_{t3}^{\mathrm{MI}}, d_{t4}^{\mathrm{MI}}, d_{t5}^{\mathrm{MI}}\right) \tag{13.11}$$

Such a model is suitable when a number of factors are considered and the difference in the weights of factors is not great (Xu et al., 2010b; Mu et al., 2014; Liu et al., 2013).

With the LO and MI membership functions of risk group t, the LO, MI, and RC of risk group t can be calculated using Eqs. (13.12)–(13.14), respectively:

$$\mathrm{LO}_{Gt} = \sum_{j=1}^{5} \left(s_j \times d_{tj}^{\mathrm{LO}}\right) \tag{13.12}$$

$$\mathrm{MI}_{Gt} = \sum_{j=1}^{5} \left(s_j \times d_{tj}^{\mathrm{LO}}\right) \tag{13.13}$$

$$\mathrm{RC}_{Gt} \sqrt{\mathrm{LO}_{Gt} \times \mathrm{MI}_{Gt}} \tag{13.14}$$

where $s_j = 1, 2, 3, 4, 5$.

Step 3: Calculate Overall LO, MI, and RC (Level 3)

To calculate the LO and MI of the overall risk (level 3), the weight of each risk group (level 2), $W_G = \{w_{G1}, w_{G2}, \ldots w_{Gq}\}$, should be determined. Here, q is the number of risk groups. The weights assigned to the LO and MI of risk group t can be calculated by Eqs. (13.15), (13.16), respectively:

$$w_{Gt}^{\mathrm{LO}} = \left(\sum_{i=1}^{k} \mathrm{LO}_i\right)_t \Big/ \sum_{t=1}^{q} \left(\sum_{i=1}^{k} \mathrm{LO}_i\right)_t \tag{13.15}$$

$$w_{Gt}^{\mathrm{MI}} = \left(\sum_{i=1}^{k} \mathrm{MI}_i\right)_t \Big/ \sum_{t=1}^{q} \left(\sum_{i=1}^{k} \mathrm{MI}_i\right)_t \tag{13.16}$$

where $\left(\sum_{i=1}^{k} \mathrm{LO}_i\right)_t$ denotes the sum of LO of k risk factors under group t; and $\left(\sum_{i=1}^{k} \mathrm{MI}_i\right)_t$ denotes the sum of MI of k risk factors under group t.

The LO and MI membership functions of the overall risk can be calculated using Eqs. (13.17)–(13.20):

$$d_{\mathrm{All}j}^{\mathrm{LO}} = \sum_{t=1}^{q} w_{Gt}^{\mathrm{LO}} \times d_{tj}^{\mathrm{LO}} \tag{13.17}$$

$$\begin{aligned}\left(D_{\mathrm{All}}^{\mathrm{LO}}\right)_{1\times 5} &= \left(W_G^{\mathrm{LO}}\right)_{1\times q} \times \left(D_G^{\mathrm{LO}}\right)_{q\times 5} \\ &= \left(d_{\mathrm{All1}}^{\mathrm{LO}}, d_{\mathrm{All2}}^{\mathrm{LO}}, d_{\mathrm{All3}}^{\mathrm{LO}}, d_{\mathrm{All4}}^{\mathrm{LO}}, d_{\mathrm{All5}}^{\mathrm{LO}}\right)\end{aligned} \tag{13.18}$$

$$d_{\mathrm{All}j}^{\mathrm{MI}} = \sum_{t=1}^{q} w_{Gt}^{\mathrm{MI}} \times d_{tj}^{\mathrm{MI}} \tag{13.19}$$

$$\begin{aligned}\left(D_{\text{All}}^{\text{MI}}\right)_{1\times 5} &= \left(W_{G}^{\text{MI}}\right)_{1\times q} \times \left(D_{G}^{\text{MI}}\right)_{q\times 5} \\ &= \left(d_{\text{All1}}^{\text{MI}}, d_{\text{All2}}^{\text{MI}}, d_{\text{All3}}^{\text{MI}}, d_{\text{All4}}^{\text{MI}}, d_{\text{All5}}^{\text{MI}}\right)\end{aligned} \tag{13.20}$$

where $(D_{G}^{\text{LO}})_{q\times 5}$ and $(D_{G}^{\text{MI}})_{q\times 5}$ are $q\times 5$ matrices that contain q matrices of $(D_{t}^{\text{LO}})_{1\times 5}$ and $(D_{t}^{\text{MI}})_{1\times 5}$, respectively.

With the overall LO and MI membership functions, the overall LO, MI, and RC can be calculated using Eqs. (13.21)–(13.23), respectively:

$$\text{LO}_{\text{All}} = \sum_{j=1}^{5}\left(s_j \times d_{\text{All}j}^{\text{LO}}\right) \tag{13.21}$$

$$\text{MI}_{\text{All}} = \sum_{j=1}^{5}\left(s_j \times d_{\text{All}j}^{\text{MI}}\right) \tag{13.22}$$

$$\text{RC}_{\text{All}} = \sqrt{\text{LO}_{\text{All}} \times \text{MI}_{\text{All}}} \tag{13.23}$$

where $s_j = 1, 2, 3, 4, 5$. Using the FSE approach, the LO and MI membership functions of each risk factor were obtained, as indicated in Table 13.3.

For example, the LO membership function of "increasing inflation rate" (R01) is calculated using Eq. (13.2):

$$\begin{aligned}\left(R_{1}^{\text{LO}}\right)_{1\times 5} &= \left(r_{11}^{\text{LO}}, r_{12}^{\text{LO}}, r_{13}^{\text{LO}}, r_{14}^{\text{LO}}, r_{15}^{\text{LO}}\right) \\ &= (0.23, 026, 0.35, 0.16, 0.00)\end{aligned}$$

Then, the LO, MI, and RC can be calculated using Eqs. (13.3)–(13.5), respectively:

$$\begin{aligned}\text{LO}_1 &= \sum_{j=1}^{5}\left(s_j \times r_{1j}^{\text{LO}}\right) \\ &= 1\times 0.23 + 2\times 0.26 + 3\times 0.35 + 4\times 0.16 + 5\times 0.00 \\ &= 2.45\end{aligned}$$

$$\begin{aligned}\text{MI}_1 &= \sum_{j=1}^{5}\left(s_j \times r_{1j}^{\text{MI}}\right) \\ &= 1\times 0.00 + 2\times 0.00 + 3\times 0.35 + 4\times 0.39 + 5\times 0.26 \\ &= 3.90\end{aligned}$$

$$\text{RC}_1 = \sqrt{\text{LO}_1 \times \text{MI}_1} = 3.09$$

To assess each risk group (level 2), the LO and MI weights of each risk factor within each risk group were calculated using Eqs. (13.6), (13.7), as shown in Table 13.4.

TABLE 13.3 LO, MI, and RC of Risk Factors Calculated Using the FSE Risk Assessment Model

			LO		MI			
Group	Risk Code	Risk Factor	*Value*	*Membership Function*	*Value*	*Membership Function*	RC	Rank
Macro-economic risk	R01	Increasing inflation rate	2.45	(0.23, 0.26, 0.35, 0.16, 0.00)	3.90	(0.00, 0.00, 0.35, 0.39, 0.26)	3.09	26
	R02	Currency exchange rate fluctuation	1.97	(0.39, 0.29, 0.29, 0.03, 0.00)	3.74	(0.00, 0.06, 0.35, 0.35, 0.23)	2.71	28
	R03	Increasing tax rate	1.87	(0.48, 0.26, 0.16, 0.10, 0.00)	4.39	(0.00, 0.00, 0.13, 0.35, 0.52)	2.86	27
Contract problems	R04	Unclear contract conditions for dispute resolution	4.00	(0.00, 0.06, 0.06, 0.68, 0.19)	4.03	(0.00, 0.03, 0.13, 0.61, 0.23)	4.02	17
	R05	Unclear contract conditions for claims and litigations	3.61	(0.00, 0.06, 0.35, 0.48, 0.10)	4.61	(0.00, 0.00, 0.10, 0.19, 0.71)	4.08	15
Client-related risk	R06	Unclear requirements of clients	4.32	(0.00, 0.03, 0.10, 0.39, 0.48)	4.55	(0.00, 0.00, 0.03, 0.39, 0.58)	4.43	4
	R07	Intervention of clients	4.19	(0.00, 0.00, 0.23, 0.35, 0.42)	4.45	(0.00, 0.00, 0.13, 0.29, 0.58)	4.32	7
	R08	Delayed payments from clients	4.23	(0.00, 0.00, 0.16, 0.45, 0.39)	4.26	(0.00, 0.00, 0.16, 0.42, 0.42)	4.24	10

Continued

TABLE 13.3 LO, MI, and RC of Risk Factors Calculated Using the FSE Risk Assessment Model—cont'd

			LO		MI			
Group	Risk Code	Risk Factor	*Value*	*Membership Function*	*Value*	*Membership Function*	RC	Rank
Design problems	R09	Unclear detailed design or specifications	4.29	(0.00, 0.10, 0.03, 0.35, 0.52)	4.74	(0.00, 0.03, 0.00, 0.16, 0.81)	4.51	3
	R10	Poor design	3.23	(0.00, 0.00, 0.77, 0.23, 0.00)	3.65	(0.00, 0.00, 0.39, 0.58, 0.03)	3.43	25
	R11	Variations in design	4.29	(0.00, 0.06, 0.16, 0.19, 0.58)	4.29	(0.00, 0.00, 0.06, 0.58, 0.35)	4.29	9
Safety risk	R12	Strict safety and health regulations	4.35	(0.00, 0.00, 0.06, 0.52, 0.42)	4.42	(0.00, 0.00, 0.13, 0.32, 0.55)	4.39	6
	R13	Construction accidents	3.71	(0.00, 0.00, 0.39, 0.52, 0.10)	4.65	(0.00, 0.00, 0.00, 0.35, 0.65)	4.15	12
Procedure complexity	R14	Complex planning approval and permit procedures	3.71	(0.00, 0.03, 0.35, 0.48, 0.13)	4.35	(0.00, 0.00, 0.16, 0.32, 0.52)	4.02	16
	R15	Delay in issuance of documents	4.39	(0.00, 0.06, 0.06, 0.29, 0.58)	4.74	(0.00, 0.03, 0.00, 0.16, 0.81)	4.56	2

Technical problems	R16	Technical complexity	3.58	(0.00, 0.00, 0.45, 0.52, 0.03)	4.06	(0.00, 0.03, 0.06, 0.71, 0.19)	3.81	20
	R17	Use of new construction method and technology	3.16	(0.00, 0.13, 0.58, 0.29, 0.00)	4.13	(0.00, 0.00, 0.26, 0.35, 0.39)	3.61	24
Human resource risk	R18	Constraints on laborer employment	3.94	(0.00, 0.00, 0.23, 0.61, 0.16)	4.48	(0.00, 0.00, 0.03, 0.45, 0.52)	4.20	11
	R19	Lack of management staff	3.97	(0.00, 0.06, 0.16, 0.52, 0.26)	3.97	(0.00, 0.10, 0.13, 0.48, 0.29)	3.97	18
	R20	Poorly trained laborers	3.48	(0.00, 0.10, 0.35, 0.52, 0.03)	4.06	(0.00, 0.00, 0.16, 0.61, 0.23)	3.76	21
Material and equipment problems	R21	Material quality problems	3.48	(0.00, 0.03, 0.58, 0.26, 0.13)	3.94	(0.00, 0.00, 0.32, 0.42, 0.26)	3.70	23
	R22	Availability of equipment and materials	3.48	(0.00, 0.00, 0.52, 0.48, 0.00)	3.94	(0.00, 0.00, 0.19, 0.68, 0.13)	3.70	23
	R23	Default supply of materials, equipment and plants	4.10	(0.00, 0.06, 0.13, 0.45, 0.35)	4.74	(0.00, 0.03, 0.03, 0.10, 0.84)	4.41	5

Continued

TABLE 13.3 LO, MI, and RC of Risk Factors Calculated Using the FSE Risk Assessment Model—cont'd

			LO		MI			
Group	Risk Code	Risk Factor	*Value*	*Membership Function*	*Value*	*Membership Function*	RC	Rank
Project team risk	R24	Project teams without relevant knowledge	3.74	(0.00, 0.03, 0.29, 0.58, 0.10)	4.19	(0.00, 0.00, 0.13, 0.55, 0.32)	3.96	19
	R25	Inefficient communication and coordination	3.90	(0.00, 0.06, 0.19, 0.52, 0.23)	4.35	(0.00, 0.03, 0.13, 0.29, 0.55)	4.12	14
	R26	Unfavorable subcontractors	3.94	(0.00, 0.03, 0.26, 0.45, 0.26)	4.35	(0.00, 0.00, 0.13, 0.39, 0.48)	4.14	13
Cost overrun risk	R27	Inaccurate cost estimation	4.65	(0.00, 0.06, 0.06, 0.03, 0.84)	4.81	(0.00, 0.00, 0.06, 0.06, 0.87)	4.73	1
	R28	Labor and materials price fluctuations	3.97	(0.00, 0.06, 0.26, 0.32, 0.35)	4.65	(0.00, 0.03, 0.03, 0.19, 0.74)	4.29	8

TABLE 13.4 Weights of Risk Factors and Groups Calculated Using the FSE Risk Assessment Model

		LO				MI			
Group	Code	*Value*	*Group Sum*	*Factor Weight*	*Group Weight*	*Value*	*Group Sum*	*Factor Weight*	*Group Weight*
Macro-economic risk	R01	2.45	6.29	0.39	0.06	3.90	12.03	0.32	0.10
	R02	1.97		0.31		3.74		0.31	
	R03	1.87		0.30		4.39		0.36	
Contract problems	R04	4.00	7.61	0.53	0.07	4.03	8.65	0.47	0.07
	R05	3.61		0.47		4.61		0.53	
Client-related risk	R06	4.32	12.74	0.34	0.12	4.55	13.26	0.34	0.11
	R07	4.19		0.33		4.45		0.34	
	R08	4.23		0.33		4.26		0.32	
Design problems	R09	4.29	11.81	0.36	0.11	4.74	12.68	0.37	0.11
	R10	3.23		0.27		3.65		0.29	
	R11	4.29		0.36		4.29		0.34	
Safety risk	R12	4.35	8.06	0.54	0.08	4.42	9.06	0.49	0.08
	R13	3.71		0.46		4.65		0.51	
Procedure complexity	R14	3.71	8.10	0.46	0.08	4.35	9.10	0.48	0.08
	R15	4.39		0.54		4.74		0.52	

Continued

TABLE 13.4 Weights of Risk Factors and Groups Calculated Using the FSE Risk Assessment Model—cont'd

		LO				MI			
Group	Code	*Value*	*Group Sum*	*Factor Weight*	*Group Weight*	*Value*	*Group Sum*	*Factor Weight*	*Group Weight*
Technical problems	R16	3.58	6.74	0.53	0.06	4.06	8.19	0.50	0.07
	R17	3.16		0.47		4.13		0.50	
Human resource risk	R18	3.94	11.39	0.35	0.11	4.48	12.52	0.36	0.10
	R19	3.97		0.35		3.97		0.32	
	R20	3.48		0.31		4.06		0.32	
Material and equipment problems	R21	3.48	11.06	0.31	0.11	3.94	12.61	0.31	0.10
	R22	3.48		0.31		3.94		0.31	
	R23	4.10		0.37		4.74		0.38	
Project team risk	R24	3.74	11.58	0.32	0.11	4.19	12.90	0.33	0.11
	R25	3.90		0.34		4.35		0.34	
	R26	3.94		0.34		4.35		0.34	
Cost overrun risk	R27	4.65	8.61	0.54	0.08	4.81	9.45	0.51	0.08
	R28	3.97		0.46		4.65		0.49	

For example, the weight assigned to the LO of "increasing inflation rate," which was one of the three risk factors ($k=3$) within risk group "macro-economic risk," was obtained using Eq. (13.6):

$$w_1^{\mathrm{LO}} = \mathrm{LO}_1 / \sum_{i=1}^{3} \mathrm{LO}_i$$
$$= 2.45/(2.45 + 1.97 + 1.87) = 2.45/6.29 = 0.39$$

The LO and MI membership functions of risk groups were also calculated for assessing these risk groups, as shown in Table 13.5.

Take "macro-economic risk" for example, its LO membership function was obtained using Eqs. (13.8), (13.9):

$$\left(D_1^{\mathrm{LO}}\right)_{1\times5} = \left(W_1^{\mathrm{LO}}\right)_{1\times3} \times \left(R_1^{\mathrm{LO}}\right)_{3\times5}$$

$$= [0.39\ \ 0.31\ \ 0.30] \times \begin{bmatrix} 0.23 & 0.26 & 0.35 & 0.16 & 0.00 \\ 0.39 & 0.29 & 0.29 & 0.03 & 0.00 \\ 0.48 & 0.26 & 0.16 & 0.10 & 0.00 \end{bmatrix}$$

$$= (0.35, 0.27, 0.28, 0.10, 0.00)$$

where $(W_1^{\mathrm{LO}})_{1\times3}$ is the LO weight matrix of this risk group, comprised of the LO weights of the three risk factors within this group; and $(R_1^{\mathrm{LO}})_{3\times5}$ is the LO membership function matrix, comprised of the LO membership functions of the three risk factors within this group.

Then, the LO of "macro-economic risk" can be calculated using Eq. (13.12):

$$\mathrm{LO}_{G1} = \sum_{j=1}^{5} \left(s_j \times d_{tj}^{\mathrm{LO}}\right)$$
$$= 1 \times 0.35 + 2 \times 0.27 + 3 \times 0.28 + 4 \times 0.10 + 5 \times 0 = 2.13$$

Similarly, the MI membership function of this risk group was calculated using Eqs. (13.10), (13.11), and the MI was obtained using Eq. (13.13). Thus, the RC of risk group "macro-economic risk" was obtained using Eq. (13.14):

$$\mathrm{RC}_{G1} = \sqrt{\mathrm{LO}_{G1} \times \mathrm{MI}_{G1}} = \sqrt{2.13 \times 4.03} = 2.93$$

The LO and MI weights of risk groups are shown in Table 13.4. The number of risk groups (k) is 11. For example, the weights assigned to the LO and MI of "macro-economic risk" were obtained using Eqs. (13.15), (13.16):

$$w_{G1}^{\mathrm{LO}} = \left(\sum_{i=1}^{3} \mathrm{LO}_i\right)_1 \Bigg/ \sum_{t=1}^{11} \left(\sum_{i=1}^{k} \mathrm{LO}_i\right)_t = 6.29/104 = 0.06$$

TABLE 13.5 LO, MI, and RC of Risk Groups and the Overall Risks Calculated Using the FSE Risk Assessment Model

	LO_G			MI_G				
Group	*Weight*	*Value*	*Membership Function*	*Weight*	*Value*	*Membership Function*	RC_G	Rank
Macro-economic risk	0.06	2.13	(0.35, 0.27, 0.28, 0.10, 0.00)	0.10	4.03	(0.00, 0.02, 0.27, 0.37, 0.34)	2.93	11
Contract problems	0.07	3.82	(0.00, 0.06, 0.22, 0.58, 0.14)	0.07	4.34	(0.00, 0.02, 0.11, 0.39, 0.48)	4.07	7
Client-related risk	0.12	4.25	(0.00, 0.01, 0.16, 0.40, 0.43)	0.11	4.42	(0.00, 0.00, 0.11, 0.37, 0.53)	4.33	2
Design problems	0.11	4.00	(0.00, 0.06, 0.27, 0.26, 0.40)	0.10	4.27	(0.00, 0.01, 0.13, 0.42, 0.43)	4.13	5
Safety risk	0.08	4.06	(0.00, 0.00, 0.21, 0.52, 0.27)	0.08	4.54	(0.00, 0.00, 0.06, 0.34, 0.60)	4.29	4
Procedure complexity	0.08	4.08	(0.00, 0.05, 0.20, 0.38, 0.37)	0.08	4.56	(0.00, 0.02, 0.08, 0.24, 0.67)	4.31	3
Technical problems	0.06	3.38	(0.00, 0.06, 0.51, 0.42, 0.02)	0.07	4.10	(0.00, 0.02, 0.16, 0.53, 0.29)	3.73	10
Human resource risk	0.11	3.81	(0.00, 0.05, 0.24, 0.55, 0.16)	0.10	4.18	(0.00, 0.03, 0.10, 0.51, 0.35)	3.99	8
Material and equipment problems	0.11	3.71	(0.00, 0.03, 0.39, 0.40, 0.17)	0.10	4.24	(0.00, 0.01, 0.17, 0.38, 0.43)	3.97	9
Project team risk	0.11	3.86	(0.00, 0.04, 0.25, 0.52, 0.19)	0.11	4.30	(0.00, 0.01, 0.13, 0.41, 0.45)	4.08	6
Cost overrun risk	0.08	4.33	(0.00, 0.06, 0.15, 0.17, 0.61)	0.08	4.73	(0.00, 0.02, 0.05, 0.13, 0.81)	4.53	1
Overall		3.83	(0.02, 0.06, 0.26, 0.40, 0.27)		4.33	(0.00, 0.01. 0.13, 0.38, 0.48)	4.07	

$$w_{Gt}^{\mathrm{MI}} = \left(\sum_{i=1}^{k} \mathrm{MI}_i\right)_t \Big/ \sum_{t=1}^{q} \left(\sum_{i=1}^{k} \mathrm{MI}_i\right)_t = 12.03/120.45 = 0.10$$

The LO and MI weights of risk groups were used to assess the overall risk level (level 3). The overall LO membership function was obtained by calculating the fuzzy composition of the risk group weight vector and the evaluation matrix using Eqs. (13.17), (13.18):

$$\left(D_{\mathrm{All}}^{\mathrm{LO}}\right)_{1\times5} = \left(W_{G}^{\mathrm{LO}}\right)_{1\times11} \times \left(D_{G}^{\mathrm{LO}}\right)_{11\times5}$$

$$= [0.06\ 0.07\ 0.12\ 0.11\ 0.08\ 0.06\ 0.11\ 0.11\ 0.11\ 0.08] \times \begin{bmatrix} 0.35 & 0.27 & 0.28 & 0.10 & 0.00 \\ 0.00 & 0.06 & 0.22 & 0.58 & 0.14 \\ 0.00 & 0.01 & 0.16 & 0.40 & 0.43 \\ 0.00 & 0.06 & 0.27 & 0.26 & 0.40 \\ 0.00 & 0.00 & 0.21 & 0.52 & 0.27 \\ 0.00 & 0.05 & 0.20 & 0.38 & 0.37 \\ 0.00 & 0.06 & 0.51 & 0.42 & 0.02 \\ 0.00 & 0.05 & 0.24 & 0.55 & 0.16 \\ 0.00 & 0.03 & 0.39 & 0.40 & 0.17 \\ 0.00 & 0.04 & 0.25 & 0.52 & 0.19 \\ 0.00 & 0.06 & 0.15 & 0.17 & 0.61 \end{bmatrix}$$

$$= (0.02, 0.06, 0.26, 0.40, 0.27)$$

Thus, the overall LO value was calculated using Eq. (13.21):

$$\mathrm{LO}_{\mathrm{All}} = \sum_{j=1}^{5} \left(s_j \times r_j^{\mathrm{LO}}\right)$$
$$= 1 \times 0.02 + 2 \times 0.06 + 3 \times 0.26 + 4 \times 0.40 + 5 \times 0.27 = 3.83$$

Similarly, the overall MI can be calculated using Eqs. (13.19), (13.20), (13.22):

$$\left(D_{\mathrm{All}}^{\mathrm{MI}}\right)_{1\times5} = \left(W_{G}^{\mathrm{MI}}\right)_{1\times11} \times \left(D_{G}^{\mathrm{MI}}\right)_{11\times5}$$
$$= (0.00, 0.01, 0.13, 0.38, 0.48)$$

$$\mathrm{MI}_{\mathrm{All}} = \sum_{j=1}^{5} \left(s_j \times r_j^{\mathrm{MI}}\right)$$
$$= 1 \times 0.00 + 2 \times 0.01 + 3 \times 0.13 + 4 \times 0.38 + 5 \times 0.48 = 4.32$$

Finally, the overall RC was obtained using Eq. (13.23):

$$\mathrm{RC}_{\mathrm{All}} = \sqrt{\mathrm{LO}_{\mathrm{All}} \times \mathrm{MI}_{\mathrm{All}}} = \sqrt{3.83 \times 4.32} = 4.07$$

Because the linguistic term mapped to 4.00 is "high," the overall RC of the green projects in Singapore was considered as high.

13.3.2 Top 10 Risk Factors of Green Construction Projects

The high overall RC confirmed that risk management was necessary for green construction in Singapore. The top 10 risk factors of green construction projects in Singapore are discussed in this section.

(1) Inaccurate cost estimation: $RC_{27}=4.73$

Among the 28 risk factors, "inaccurate cost estimation" was ranked top, mainly because of its high LO and MI values. Cost has been considered as one of the most important factors to contractors (Hlaing et al., 2008) and with the introduction of green features, there is more room for errors. This is supported by Robichaud and Anantatmula (2010) who argued that the sustainable building product market was relatively immature given the youth of the green construction industry and that there had been no particular company standing out in green building materials (McGraw-Hill Construction, 2006a). Additionally, expensive technologies, products and materials, and consultancy services associated with green buildings are likely to take a toll on the construction budget (Hwang and Tan, 2012). As a result, there is great difficulty in accurately estimating costs and any inaccurate estimate poses a critical risk to green projects.

(2) Delay in issuance of documents: $RC15=4.56$

"Delay in issuance of documents" occupied the second position in green projects. Late issuance of relevant documents is likely to result in project delay (Hwang et al., 2013c), especially when it occurs on the critical path. If a change order is not issued on time and the construction activities continue, rework will occur and cost will increase. Thus, this risk factor was rated high.

(3) Unclear detailed design or specifications: $RC9=4.51$

"Unclear detailed design or specifications" was ranked third, attributed to its high LO and MI scores. Green projects typically incorporate more advanced and intricate systems of interacting elements (Kubba, 2010). During design, the impact of the elements on each system should be considered as a whole. If the architect developed the design before the design engineers were selected, little attention would be given to the mechanical, electrical, and hydraulic services of the building (Love et al., 1999). Thus, design engineers and contractor engineers would not be clear about the design, leading to construction errors and rework.

(4) Unclear requirements of clients: $RC6=4.43$

"Unclear requirements of clients" received the fourth position and was considered to pose a high risk to contractors in green projects. In green construction, due to the complexity involved, clients may not be able to give clear requirements and instructions, thus increasing the difficulty to contractors of carrying out projects.

(5) Default supply of materials, equipment and plants: RC23 = 4.41

"Default supply of materials, equipment and plants" was ranked fifth in green projects. This risk factor tends to lead to the unavailability of materials, equipment and plants, and project delay. In green projects, some materials, equipment, or plants specifically for green building are likely to be imported from overseas, requiring several weeks or months to be delivered on site. In addition, few alternative supply sources make replacement difficult should a delay in shipment occur. Thus, any technical hitch on the delivery of imported materials, especially those concerning the critical activities in a schedule, would adversely impact project schedule. This result was consistent with the finding of Hwang et al. (2015d) that late delivery of materials and equipment was one of the most critical causes of delay in green projects in Singapore.

(6) Strict safety and health regulations: RC12 = 4.39

"Strict safety and health regulations" received the sixth position. Workers in green projects face new, high-risk tasks. Specifically, Fortunato III et al. (2012) found that workers in green projects suffered a 24% increase in falls to lower levels during roof work when installing solar panels, experienced a 19% increase in eyestrain when installing reflective roof membranes, and faced a 14% increase in exposure to harmful substances when installing innovative wastewater technologies. The Ministry of Manpower (MOM, 2006) of Singapore has issued the Workplace Safety and Health (Risk Management) Regulations. All the contractors should comply with these regulations and assess the safety and health risks.

(7) Intervention of clients: RC7 = 4.32

Another risk factor from clients was "intervention of clients," ranked seventh, implying that improper intervention of clients was not uncommon in green projects. This result was consistent with the argument of Durmus-Pedini and Ashuri (2010) that clients may not be willing to leave their comfort zone in dealing with new team members and with new technologies and processes, and that this may increase the probability of occurrence of intervention.

(8) Labor and materials price fluctuations: RC28 = 4.29

"Labor and materials price fluctuations" occupied the eighth position. Price fluctuations have been active due to a sharp fluctuation in oil and steel prices worldwide (Skorupka, 2008). Green materials, which are inevitably used in green projects, are generally more expensive. Because an increased material price would result in lower profitability for the contractors, this risk was rated high by project managers.

(9) Variations in design: RC11 = 4.29

"Variations in design" was another critical risk. Clients, especially private clients, tend to change designs according to the changing economic climate, to meet customer needs, or for marketing reasons. Variations in design impact the plans of contractors and may even require extensive redesign (Ogunlana et al., 1996). In Singapore, variations in design have been found to significantly contribute to rework and delay in construction projects (Arain and Low, 2006), and client-initiated variations was one of the most critical causes of delay in green projects (Hwang et al., 2015d).

(10) Delayed payments from clients: RC8 = 4.24

"Delayed payments from clients" received the tenth position. This risk factor is usually attributed to the financial difficulties of clients because financing by clients ensures the progress payments of completed work (Hwang et al., 2013c). Delayed payments from clients would result in poor financial status of contractors, project delays or even shutdown (Long et al., 2008). In contrast, projects tend to achieve better performance than desired when payments to the contractor were released promptly (Iyer and Jha, 2006).

13.3.3 Top Three Risk Groups of Green Construction Projects

(1) Cost overrun risk: $RC_{G11} = 4.53$

Among the 11 risk groups, "cost overrun risk" was ranked top, suggesting that cost overrun was the most critical problem in green projects. In the context of green construction, cost overrun may make green construction unattractive because a client may consider that the green features of the building have led to the cost overrun. In addition, cash flow is the lifeline for business continuity but the high upfront costs associated with green construction are likely to threaten the achievement of the cost objectives of green projects. Although Singapore's government provides various financial incentive schemes, such as the Green Mark Incentive Scheme for Existing Buildings (GMIS-EB) and the Building Retrofit Energy Efficiency Financing (BREEF) schemes, these would not significantly assist the building owners financially because of the uncertainty in receiving the cash incentives, which greatly increases the difficulty in accurately estimating the costs of green projects.

(2) Client-related risk: RCG3 = 4.33

"Client-related risk" received the second position, implying that green projects were plagued with client-related problems. All the risk factors within this group were ranked within the top 10. Unclear requirements, intervention, and late payment were the typical problems related to clients. This result echoed Hwang et al. (2014), who reported that the problems of clients were the most importance causes of rework and low productivity in Singapore. Additionally, unclear requirements of clients usually lead to designers' misinterpretation of the requirements, which was recognized as a critical factor influencing green project schedule performance in Singapore (Hwang et al., 2015d).

(3) Procedure complexity: RCG6 = 4.31

This risk group was ranked third. Green construction projects tend to involve higher-level complexity, which results in more complex planning approval and permit procedures (Greenwald, 2012). Also, the procedure complexity may result in delay in planning approval and permit, thus possibly leading to poor schedule performance of green projects.

It merits attention that "macro-economic risk" ($RC_{G1} = 2.93$) was the least critical group but it still had a high MI value, suggesting that once this risk occurred, its impact would be high. The low LO value resulted from the perception that Singapore has a relatively stable macro-economic environment. In addition, the low RC of this risk group was attributed to the low RC of its three risk factors, which were ranked the last three in the risk ranking, as shown in Table 13.3.

13.4 SUMMARY

Green construction projects are inevitably plagued with risks and this chapter attempts to assess the risks associated with green construction projects. As fuzzy set theory can handle the problems related to ambiguous, subjective and imprecise judgments, which are inevitably involved in risk assessment, it was employed to assist in developing the risk assessment model. Categorizing 28 risk factors into 11 groups, a questionnaire was developed and disseminated to project managers in Singapore, receiving 31 valid responses. Using the proposed risk assessment model and the data from the survey, the likelihoods of occurrence, magnitudes of impact, and risk criticalities of the risk factors, groups, and the overall risk were obtained. The results indicated that "inaccurate cost estimation," "delay in issuance of documents," and "unclear detailed design or specifications" were the top three critical risk factors in green construction projects. At the risk group level, "cost overrun risk," "client-related risk," and "procedure complexity" were the top three. In addition, the overall risk criticality calculated by the assessment model was high, indicating that risk management is still necessary for green construction in Singapore.

Chapter 14

Risk Management for Green Retrofit, Residential, and Commercial Construction Projects

14.1 INTRODUCTION

Risk management, as a permanent and classic topic in construction engineering and management (CEM) research, has been addressed by a considerable research effort for decades (Edwards and Bowen, 1998). Research interests in this field range from identification (e.g., Zou et al., 2007; Ke et al., 2011), allocation (e.g., Jin, 2010; Ng and Loosemore, 2007), and assessment (e.g., Xu et al., 2010b; Dikmen et al., 2012), to the investigation of specific risks, such as safety (e.g., Hallowell and Gambatese, 2009), cost (e.g., Khodakarami and Abdi, 2014), schedule (e.g., Ökmen and Öztaş, 2008), political (e.g., Deng et al., 2014b), and legal risks (e.g., Ling and Hoang, 2009). Recently, because of the rapid development in green construction projects, academia has started to transfer its attention to risk management in green construction projects. For instance, Fortunato et al. (2012) identified and evaluated design elements and construction management practices that may induce safety and health risks in green building projects. Dewlaney et al. (2012) quantified the percent increase of safety risk in a project that applies some particular design strategies and construction methods to earn a green certificate. Yang and Zou (2014) and Yang et al. (2016) assessed and analyzed risks among stakeholders in complex green construction projects from the perspective of their interactions. However, current literature shows that little research effort has been made to systematically examine the risk factors in green retrofit, residential, and commercial construction projects.

Green retrofit is a kind of upgrade to an existing building that can reap a wide range of benefits, including operational cost savings, quality assurance, better indoor air and environmental quality, and improved employee productivity (Deloitte, 2008; McGraw-Hill Construction, 2013). In Singapore, existing buildings consume a third of Singapore's total end-use electricity

Performance and Improvement of Green Construction Projects. https://doi.org/10.1016/B978-0-12-815483-0.00014-4

(BCA, 2010b) and thus, special emphasis has been placed on green retrofit to existing buildings in order to improve their energy and environmental performance. For instance, the Green Mark Incentive Scheme for Existing Buildings launched by the authority can fund up to 35% of the cost of a green retrofit project. The efficacy of green retrofit has also been proved. In Singapore, it has been recognized that green retrofit can decrease operating costs by 14% over 5 years, with a payback time of approximately 6 years for green investment (McGraw-Hill Construction, 2013), and achieve an average of 20% in energy savings (BCA, 2014a).

In a typical densely populated city-state like Singapore, a large number of residential buildings must be built to address the people's housing needs (Agarwal et al., 2016). Based on the statistics revealed by the BCA, residential buildings have grown into the largest piece of the local construction market over recent years (BCA, 2016f). Thus, naturally residential buildings have become the primary target for the local construction authority to promote green buildings. Recently, local authorities like the Housing and Development Board (HDB) of Singapore and the BCA have devoted considerable effort to develop green residential buildings in the country. For instance, in 2007 the HDB started to develop environmentally-friendly public housing blocks (e.g., Punggol Eco-Town). In 2010, the BCA launched the BCA Green Mark for New Residential Buildings to encourage developers and building owners and architecture firms to develop green and sustainable residential buildings (BCA, 2010d). Thus, it is necessary to investigate the various risks involved in these green residential construction projects.

As a specific type of building, commercial buildings (e.g., office buildings, malls, retail stores, hotels, medical centers, and warehouses) play an important role in serving people's daily life (Yik et al., 2001; Ling and Tiong, 2008). However, commercial buildings are also criticized as inefficient energy consumers. According to BCA (2014d), about 38% of electricity in Singapore is consumed by commercial buildings. Given this distinction, the commercially built structures in Singapore have been urged to change their construction practices and go sustainable in recent years (Lee et al., 2015). In fact, the Singaporean construction society is expecting more green commercial construction projects. According to the newly published World Green Building Trends 2016, 53% of Singapore-based respondents expect to develop new commercial building projects in the next 3 years, which is above the global average of 46% (Dodge Data & Analytics, 2016).

The objectives of this chapter are (1) to identify the critical risks in green retrofit, residential and commercial construction projects, (2) to compare their risk criticalities with their counterparts in conventional retrofit, residential and commercial construction projects, and (3) to propose relevant risk mitigation measures that can tackle these critical risks.

14.2 METHOD AND DATA PRESENTATION

To examine the risks in green retrofit construction projects, a comprehensive literature review was conducted first, which generated an initial list of risks as shown in Table 14.1. Subsequently, a questionnaire was developed based on the literature review results and was then sent to three professionals certified with either Green Mark Professional or Green Mark Manager to solicit their comments on the readability, comprehensiveness, and accuracy of the questionnaire. After that, a total of 145 finalized questionnaires were randomly disseminated to all the BCA-certified Green Mark Professionals and Green Mark Managers in Singapore to gather their opinions about the risks in green retrofit projects in terms of their likelihood of occurrence (LO) and magnitude of impact (MI) with the aid of the two five-point Likert scales shown in Table 14.2. Finally, 30 valid responses were received, representing a response rate of 21%, which was consistent with the norm of 20%–30% with most questionnaire surveys of the construction industry (Akintoye, 2000). The profiles of the respondents and their companies are provided in Table 14.3. These respondents have an average of 13.2 years of experience in retrofit projects and an average of 4.4 years in green retrofit projects. 53% of the respondents possessed over 10 years of experience in retrofits, whereas 63% had more than 4 years of experience in green retrofits, which was reasonable because the 2nd Green Building Masterplan focusing on green retrofits was launched in 2009. In addition, the respondents were from 10 consultancy firms (33%), six developers (20%), six facility management firms (20%), five ESCOs (17%), and three construction firms (10%).

A similar research strategy was applied in the investigation of risks in green residential construction projects. First, a comprehensive literature review was conducted, which identified 42 risks that may occur during the implementation of green residential construction projects (as shown in Table 14.4). Subsequently, the 42 identified risks formed a questionnaire which was reviewed by two industry experts who had at least 5 years of experience in both conventional and green residential construction projects, aiming to improve the statement comprehensiveness, readability, and accuracy. After that, the finalized questionnaires were sent to 100 local construction companies that have experience in both green and conventional residential construction projects in Singapore, from which a total of 30 valid responses were received. The profiles of the respondents and their companies are presented in Table 14.5. It should be noted that the respondent companies comprised various project stakeholders, such as consultants, developers, contractors, and architecture firms. Moreover, it is noteworthy that 53% of the respondents have at least 5 years of experience in traditional

TABLE 14.1 Risks Associated With Green Retrofit Projects

Code	Risks Factors	References																	
		1	*2*	*3*	*4*	*5*	*6*	*7*	*8*	*9*	*10*	*11*	*12*	*13*	*14*	*15*	*16*	*17*	*18*
R01	Financial risk	√	√	√	√														
R02	Return on investment (ROI) risk	√				√													
R03	Energy saving uncertainty			√		√	√	√											
R04	Accreditation of energy service companies (ESCOs)								√										
R05	Regulatory risk					√				√									
R06	Legislative risk	√																	
R07	Warranty risk										√	√							
R08	Market risk	√				√													
R09	Industry risk	√											√						
R10	Delay in project completion													√	√	√			
R11	Productivity risk												√						

R12	Manpower supply and availability												√	√					
R13	Material supply and availability									√							√		
R14	Team performance risk	√																	
R15	Reliability and accuracy of benchmark								√										
R16	Quality risk				√					√									
R17	Preretrofit tenants' cooperation risk					√										√			
R18	Postretrofit tenants' cooperation risk					√													
R19	Different concerns of stakeholders	√																√	
R20	Lack of streamlined tools and processes					√													√

References: 1. Durmus-Pedini and Ashuri (2010); 2. WorldGBC (2013); 3. Penny (2012); 4. Zurich Services Corporation (2011); 5. Lockwood (2009); 6. Winston (2009); 7. Yudelson (2010b); 8. UNEP (2009); 9. Moresco (2009); 10. CBRE (2011); 11. ICF International (2007); 12. Drew (2011); 13. Pitt (2013); 14. Chanter and Swallow (2007); 15. Sanvido and Riggs (1991); 16. Davis Langdon Inc. (2007); 17. Berardi (2013); 18. Hwang and Tan (2012).

TABLE 14.2 Rating Scales for LO and MI

Likelihood of Occurrence			Magnitude of Impact	
LO	*Linguistic Terms*	*Likelihood of References (%)*	*MI*	*Linguistic Terms*
1	Rarely	<20	1	Very small
2	Somewhat likely	20–40	2	Small
3	Likely	40–60	3	Medium
4	Very likely	60–80	4	Large
5	Almost definite	>80	5	Very large

TABLE 14.3 Profile of Respondents for the Questionnaire Survey Regarding Green Retrofit Projects

Characteristics		Frequency	Percentage
Experience in retrofitting	<5 years	2	6.67
	6–10 years	12	40.00
	11–15 years	6	20.00
	15–20 years	7	23.33
	>20 years	3	10.00
	Mean	*13.2 years*	
Experience in green retrofitting	2–3 years	11	36.67
	4–5 years	12	40.00
	6–7 years	4	13.33
	>8 years	3	10.00
	Mean	*4.4 years*	
Firm type	Consultancy	10	33.33
	Developer	6	20.00
	Facility management	6	20.00
	ESCO	5	16.67
	Construction firms	3	10.00

TABLE 14.4 Risks Associated With Green Residential Construction Projects

Code	Risks	Source									
		[1]	*[2]*	*[3]*	*[4]*	*[5]*	*[6]*	*[7]*	*[8]*	*[9]*	*[10]*
R1	Green building policies change								✓		
R2	Energy saving uncertainty						✓	✓		✓	
R3	Inflation					✓					✓
R4	Fluctuation in exchange rates					✓		✓			✓
R5	High crime rate										✓
R6	Complex procedures to obtain approvals					✓			✓		✓
R7	Safety and health		✓	✓							✓
R8	Employment constraint					✓	✓	✓	✓	✓	✓
R9	Pollution restrictions						✓			✓	
R10	Import/export restrictions					✓		✓			
R11	Unclear contract conditions for claims and litigations										✓
R12	Unclear contract conditions for dispute resolution							✓			✓
R13	Unclear allocation of roles and responsibilities					✓	✓		✓	✓	
R14	Shortage of funds	✓					✓	✓	✓	✓	
R15	Unclear Requirements					✓	✓		✓	✓	✓

Continued

TABLE 14.4 Risks Associated With Green Residential Construction Projects—cont'd

Code	Risks	Source									
		[1]	[2]	[3]	[4]	[5]	[6]	[7]	[8]	[9]	[10]
R16	Inappropriate interventions of clients										✓
R17	Loose control over subcontractors							✓			
R18	Warranties to homeowners of green building				✓			✓			
R19	Unclear design details and specifications				✓	✓	✓		✓	✓	✓
R20	Being fined for failing to achieve Green Mark standards				✓				✓		
R21	Poor communication among projects stakeholders					✓	✓		✓	✓	
R22	Lack of qualified professionals with proper design expertise					✓	✓	✓	✓	✓	
R23	Claims of overstated or unverifiable benefits				✓						
R24	Unfamiliarity of job requirement					✓	✓		✓	✓	
R25	Exposed to lawsuit for failing to achieve GM standards				✓				✓		
R26	Overlooked initial high cost										✓
R27	Technical Issues					✓	✓		✓	✓	✓
R28	Lack of availability of green materials and equipment					✓	✓	✓	✓	✓	✓

R29	Lack of technical expertise					✓	✓	✓	✓	✓	
R30	Unfamiliarity with green materials and construction technologies					✓	✓	✓	✓	✓	✓
R31	Lack of experience					✓	✓	✓	✓	✓	
R32	Setting expectations too high							✓			
R33	Unskilled workers					✓		✓			✓
R34	Poor Design				✓	✓	✓		✓	✓	✓
R35	Poor Workmanship				✓	✓	✓	✓	✓	✓	
R36	Inefficient Communication					✓	✓		✓	✓	✓
R37	Lack of management staff										✓
R38	Unfamiliarity with construction process					✓	✓	✓		✓	
R39	Inaccurate estimation					✓		✓	✓		✓
R40	Unfamiliarity with new technology rates					✓	✓	✓		✓	
R41	Fluctuations in labor/ material rates					✓		✓	✓		✓
R42	High Target for Green Mark Rating								✓		

Note: [1] Ranaweera and Crawford (2010); [2] Dewlaney et al. (2011); [3] Fortunato III et al. (2011); [4] Tollin (2011); [5] Zou and Couani (2012); [6] Yang and Zou (2014); [7] Hwang et al. (2015a); [8] Qin et al. (2016); [9] Yang et al. (2016); and [10] Zhao et al. (2016).

TABLE 14.5 Profile of Respondents for Green Residential Projects

Profile	Frequency	Percentage
Company (total = 30)		
Type		
Consultant	12	40.00
Developer	1	3.33
Contractor	10	33.33
Architect	7	23.33
Respondent (total = 30)		
Job title		
Developer	1	3.33
Architect	6	20.00
Contractor	3	10.00
Quantity surveyor	15	50.00
Consultant	5	16.67
Years of experience in traditional residential building projects		
Less than 1 year	2	6.67
1–2 years	4	13.33
3–4 years	8	26.67
5–10 years	11	36.67
More than 10 years	5	16.67
Years of experience in green residential building projects		
Less than 1 year	3	10.00
1–2 years	8	26.67
3–4 years	12	40.00
5–10 years	7	23.33
More than 10 years	0	0.00

residential construction projects, and 63% of the respondents have at least 3 years of experience in green residential construction projects. All of this suggests that the respondents have the requisite experience to address the research questions of the survey and that the collected data should be representative.

Regarding the research methods for investigating the risks in green commercial construction projects, they are same as those adopted in examining risks in green retrofit and residential construction projects. First, a comprehensive literature review was conducted, which identified a series of risks that may occur during the implementation of green commercial construction projects. Subsequently, interviews were carried out with three experienced industry experts to fit the identified risks into the context of Singapore. Then, the finalized risks (as listed in Table 14.6) formed a questionnaire that was disseminated to 150 professionals who were certified either as Green Mark Managers or Green Mark Professionals. Finally, a total of 27 valid responses were received. Table 14.7 presents the profiles of the respondents and their companies. It should be noted that all the respondents possess at least 5 years of experience in conventional commercial construction projects, and 77.78% of respondents possess at least 5 years of experience in green commercial construction projects, suggesting that the respondent panel has the requisite experience to address the research questions.

TABLE 14.6 Risks Associated With Green Commercial Construction Projects

Category	Code	Risk Factor	Source
Technical risk	A1	Alteration and variation with the design during construction	[1], [5], [12], [17], [18], [22]
	A2	Insufficient or incorrect design information	[1], [4], [5], [7], [10], [12], [16], [17], [22], [25]
	A3	Improper quality control and defective work	[1], [2], [4], [9], [12], [15], [19], [22], [25], [26]
	A4	Poor definition of scope and change in scope	[1], [3], [4], [5], [6], [12], [15], [20], [22], [25]
	A5	Selection of construction technique and sequence	[4], [7], [10], [12], [14], [16], [17], [18], [22]
	A6	Material suitability and accessibility and shortage of green material	[28]
	A7	Durability of green materials (new technology)	[28]

Continued

TABLE 14.6 Risks Associated With Green Commercial Construction Projects—cont'd

Category	Code	Risk Factor	Source
Labor risk	B1	Injuries and accidents	[1], [2], [4], [5], [6], [9], [12], [13], [19], [20], [22], [25]
	B2	Damages caused by human error	[11]
	B3	Shortage of skills or techniques; lack in technical modern knowledge	[4], [5], [6], [7], [10], [12], [13], [15], [18], [20], [22], [23], [25]
	B4	Lack of motivation and productivity to change from their traditional practices	[28]
Management risk	C1	Inadequate insurance	[2], [3], [5], [22]
	C2	Poor communications among stakeholders	[3], [4], [6], [7], [9], [12], [14], [16], [19], [20], [22], [24], [25], [26]
	C3	Improper project feasibility and planning	[4], [5], [6], [9], [12], [15], [17], [22], [25]
	C4	Customer expectation management	[28]
	C5	Difficulty in comprehending green specifications in contract details	[28]
	C6	Limited availability and reliability of green suppliers	[28]
	C7	Resistance from stakeholders/client to adopt green ideas	[28]
	C8	Special request from client about specified green technologies to be used	[28]
	C9	Delay caused by the frequent meetings with green specialists	[28]

TABLE 14.6 Risks Associated With Green Commercial Construction Projects—cont'd

Category	Code	Risk Factor	Source
Financial risk	D1	Inflation (labor and material cost increase)	[1], [2], [4], [5], [6], [7], [8], [10], [11], [12], [14], [15], [16], [17], [19], [20], [21], [22]
	D2	Financial strength of project stakeholder	[1], [2], [4], [5], [6], [7], [10], [11], [12], [13], [14], [16], [17], [20], [21], [24], [26], and [27]
	D3	Inexact prediction of market demand	[4], [6], [7], [11], [13], [14], [15], [16], [17], [20], [26]
	D4	Currency and interest rate volatility worsened by the import of green materials	[28]
	D5	Difficulties in project budgeting due to unfamiliarity in green projects	[28]
	D6	Additional costs due to green material and equipment	[28]
Legal risk	E1	Breach of contract (i.e., claims and disputes)	[11], [12], [14], [15], [18], [19], [22], [25]
	E2	Intellectual property protection (e.g., special design/construction method)	[6], [22]
	E3	Delay/more time required for approval	[2], [5], [6], [7], [10], [12], [14], [16], [19], [20], [25]
	E4	Lack of documents and information for new green technologies	[28]
Environmental Risk	F1	Changes in weather	[2], [7], [14], [15], [16], [19], [20], [22], [25]
	F2	Unforeseen adverse site conditions and geographical conditions	[1], [2], [4], [5], [7], [9], [10], [12], [14], [15], [16], [18], [19], [20], [22], [25]
	F3	Air, sound and water pollution	[1], [5], [12], [15], [16], [21], [25]

Continued

TABLE 14.6 Risks Associated With Green Commercial Construction Projects—cont'd

Category	Code	Risk Factor	Source
Political risk	G1	Corruption and bribery	[8], [11], [12], [14], [16], [18], [19], [20], [22], [25], [27]
	G2	Change in local regulations/change in governmental policies	[1], [2], [4], [6], [7], [10], [11], [12], [13], [14], [15], [16], [17], [19], [20], [21], [23], [24], [25], [26], [27]

[a]The critical risk factor was excluded due to its irrelevance to green commercial construction projects in Singapore.
Source of reference: [1] ANDI (2006); [2] Ling and Hoi (2006); [3] Gruneberg et al. (2007); [4] Tang et al. (2007); [5] Zou et al. (2007); [6] Jha and Devaya (2008); [7] Roumboutsos and Anagnostopoulos (2008); [8] Skorupka (2008); [9] Le et al. (2009); [10] Iyer and Sagheer (2010); [11] Wibowo and Mohamed (2010); [12] Zou and Li (2010); [13] Aritua et al. (2011); [14] Chan et al. (2011a); [15] Chan et al. (2011b); [16] Ke et al. (2011); [17] Marques and Berg (2011); [18] Rebeiz (2012); [19] Chuing Loo et al. (2013); [20] Hwang et al. (2013b); [21] Zhao et al. (2013a); [22] Al-Sabah et al. (2014); [23] Chien et al. (2014); [24] Yang and Zou (2014); [25] El-Sayegh and Mansour (2015); [26] Hwang et al. (2015a); [27] Wang et al. (2015); and [28] interview supplementary.

TABLE 14.7 Profile of Respondents Attending the Green Commercial Construction Project Survey

Profile	Frequency	Percentage
Respondent (total = 27)		
Designation		
Project manager	12	44.44
Construction engineer	7	25.93
Architect	6	22.22
Researcher (Energy Company)	2	7.41
Years of experience in the traditional commercial construction projects		
<5 years	Nil	0.00
5–10 years	6	22.22
11–15 years	8	29.63
16–20 years	7	25.93
21–30 years	5	18.52
>30 years	1	3.70

TABLE 14.7 Profile of Respondents Attending the Green Commercial Construction Project Survey—cont'd

Profile	Frequency	Percentage
Years of experience in the green commercial construction projects		
<5 years	6	22.22
5–10 years	18	66.67
11–15 years	3	11.11
16–20 years	0	0.00
21–30 years	0	0.00
>30 years	0	0.00
Company (total = 25)		
Type		
Project management firm	7	25.90
Architectural consultancy	5	18.50
Construction firm	13	48.1
Energy policy research	2	7.4
Years of experience in the construction industry		
<5 years	2	8.00
5–10 years	15	60.00
>10 years	8	32.00
Years of experience in the green commercial construction projects		
<5 years	7	28.00
5–10 years	10	40.00
>10 years	8	32.00
Project		
Traditional construction projects (total = 293)		
<SGD15 million	44	15.02
SGD15–SGD50 million	131	44.71
>SGD50 million	118	40.27
Green commercial building projects (total = 30)		
<SGD15 million	0	0
SGD15–SGD50 million	2	6.67
>SGD50 million	28	93.33

14.3 ANALYSIS RESULTS AND DISCUSSIONS: GREEN RETROFIT CONSTRUCTION PROJECTS

14.3.1 Risk Ranking

Cronbach's alpha is commonly used as a measure of the internal consistency or reliability of data. Alpha should be at least 0.7 for a scale to be reliable (Nunnally, 1978). For this survey, the alpha values of LO and MI in green retrofitting were 0.84 and 0.88, respectively. Thus, the survey data were reliable. Using Eqs. (3.1), (3.2), the LO, MI, and RC values of each risk were calculated and they are presented in Table 14.8. In terms of LO values, the five risks most likely to occur were: "postretrofit tenants' cooperation risk," "market risk," "regulatory risk," "material supply and availability," and "different concerns of stakeholders." Additionally, six risks received LO values below 3.00, indicating that they are less likely to occur in the context of Singapore.

As for the MI values, the six most impactful risks were: "financial risk," "energy saving uncertainty," "postretrofit tenants' cooperation risk," "regulatory risk," and "preretrofit tenants' cooperation risk." It merits attention that all the 20 risks received MI values above 3.00, and eight of them obtained MI values above 4.00.

The risks were ranked according to the RC values. The risks with RC above 15.00 are discussed below. In the risk ranking, "postretrofit tenants' cooperation risk" occupied the top position (RC = 18.97), attributed to its top LO and third highest MI. This result suggests that tenants are likely to be uncooperative after green retrofit completion. Without tenants' cooperation, energy-saving technology will not be used at the best efficiency and all the potential benefits would not be secured. Consequently, the owner would not have sufficient evidence to demonstrate the energy savings to the BCA, and may not be eligible for the financial incentives under the GMIS-EBP or BREEF Scheme. In addition, the BCA re-assesses certified Green Mark buildings every 3 years (BCA, 2014b). The lack of tenants' cooperation would make the building fail in the Green Mark re-assessment.

"Regulatory risk" was ranked second (RC = 17.77), attributed to its high LO and MI values. In Singapore, the regulatory environment relating to green building has been evolving in the past 10 years, thus causing the respondents to perceive this risk as highly likely to occur and impactful. The Green Mark Scheme was voluntary when it was launched in 2005, and became mandatory when the Building Control (Environmental Sustainability) Regulations were enacted in 2008. After December 2010, all new buildings had to comply with a higher Green Mark standard. This required an additional 10% in energy savings and the minimum standard was also 28% higher than that released in 2005. In addition, from the second half of 2013, building owners were required to fulfill three prerequisites under the Building Control Act: (1) achieve minimum Green Mark standard for existing buildings when a cooling system is installed or retrofitted; (2) carry out 3-yearly energy audits on

TABLE 14.8 LO, MI, and RC of Risks in Green Retrofit Projects in Singapore

Code	Risks	LO		MI		RC	
		Mean	*Rank*	*Mean*	*Rank*	*Mean*	*Rank*
R01	Financial risk	3.80	6	4.37	1	16.83	4
R02	ROI risk	3.53	12	3.73	16	13.50	11
R03	Energy saving uncertainty	2.73	18	4.33	2	11.83	15
R04	Accreditation of ESCOs	2.73	19	3.87	12	10.73	16
R05	Regulatory risk	4.13	3	4.27	4	17.77	2
R06	Legislative risk	2.57	20	3.87	13	10.07	18
R07	Warranty risk	2.97	15	3.47	18	10.13	17
R08	Market risk	4.17	2	4.03	7	17.00	3
R09	Industry risk	3.30	13	3.93	10	13.00	13
R10	Delay in project completion	3.30	14	3.77	15	12.17	14
R11	Productivity risk	2.87	17	3.20	20	9.43	20
R12	Manpower supply and availability	3.63	10	3.67	17	13.30	12
R13	Material supply and availability	3.97	4	3.90	11	15.30	7
R14	Team performance risk	3.57	11	4.03	7	14.40	10
R15	Reliability and accuracy of benchmark	3.80	7	3.80	14	14.47	9
R16	Quality risk	3.70	9	4.10	6	15.20	8

Continued

TABLE 14.8 LO, MI, and RC of Risks in Green Retrofit Projects in Singapore—cont'd

Code	Risks	LO		MI		RC	
		Mean	*Rank*	*Mean*	*Rank*	*Mean*	*Rank*
R17	Preretrofit tenants' cooperation risk	3.80	8	4.17	5	15.90	5
R18	Postretrofit tenants' cooperation risk	4.40	1	4.30	3	18.97	1
R19	Different concerns of stakeholders	3.87	5	3.97	9	15.37	6
R20	Lack of streamlined tools and processes	2.97	16	3.30	19	9.93	19

building cooling systems; and (3) submit building information and energy consumption data annually (BCA, 2012).

"Market risk" received the third position (RC = 17.00). This risk obtained the second highest LO, indicating that it was very likely to occur. The rationale behind this high RC was collected from the postsurvey interviews. In the property market, most practitioners lack relevant knowledge of green technologies and may find it difficult to clearly express the possible benefits to clients. Hence, the owner was not likely to receive the expected market value. Also, many tenants tended not to associate green technologies with improved quality of life, and believed that the costs of green retrofits were transferred to their rent.

"Financial risk" was ranked fourth (RC = 16.83), attributed to its top MI value. The cash flow is the lifeline for business continuity but the high upfront costs associated with green retrofits are likely to threaten the achievement of the cost objectives of green retrofit projects (Durmus-Pedini and Ashuri, 2010; Zurich Services Corporation, 2011). Although Singapore's government provided various financial incentive schemes, such as the GMIS-EB, the GMIS-EBP and the BREEF Schemes, they do not significantly assist the building owners financially because of the uncertainty in receiving the cash incentives. In addition, the cash incentive in the GMIS-EB, which aimed to co-fund up to 50% (capped at $3 million) of the costs of supply and installation of energy efficient equipment and professional services, has expired with effect from 28 April 2014 because the funds for the cash incentives have been fully committed (BCA, 2013a). Moreover, the S$50 million GMIS-EBP is only applicable to SME tenants and building owners and building owners with at least 30% of their tenants who are SMEs.

"Preretrofit tenants' cooperation risk" was the fifth most critical risk (RC = 15.90), possibly because of its high MI value. This risk was perceived as impactful because green retrofits are usually conducted with the ongoing operation of the tenants. The postsurvey interviewees indicated that it was a challenge to obtain the tenants' cooperation because most of them believed that green retrofitting was not necessary and the work inevitably impacted their business. These findings were consistent with the viewpoint of Miller and Buys (2008) that tenants tended to view retrofitting work as disruptive.

"Different concerns of stakeholders" was ranked sixth (RC = 15.37), suggesting that different or conflicting concerns of stakeholders had significant influence on green retrofit projects in Singapore. It was not surprising that different stakeholders held different concerns because of the fragmentary nature of the construction industry. As an interviewee indicated, the building owner and tenants usually have different concerns when the building is upgraded or retrofitted. The building owner has to comply with the Building Control Act and ensure that the building is at least certified by the Green Mark. The owner would also require the cooperation of the tenants to collect evidence demonstrating energy saving and also to apply for financial incentives from the government. In contrast, the tenants may focus on the disturbance caused by the retrofit work on their business and operations. This finding was consistent with the argument of Berardi (2013) that the disparity in stakeholders' concerns hindered the adoption of green technologies.

"Material supply and availability" received the seventh position (RC = 15.37) with a high LO. It is common that adequate material supply is the precondition of nearly all construction activities, and a shortage of material leads to project shutdown and delay (Ogunlana et al., 1996). In Singapore, the construction material supply, including the green material supply, greatly depends on imports (Hwang and Leong, 2013). Thus, any foreign restriction on exports would cause material unavailability, impacting project objectives.

"Quality risk" was ranked eighth (RC = 15.20) with a high MI score. Quality objective is a common objective of all construction projects. In green retrofit projects, the quality of the energy-saving equipment is of great importance because it determines whether a building is really green according to the Green Mark Scheme. Inferior quality of the equipment might make the energy efficiency of a building similar to that of a conventional building, and thus the green retrofitting would not make any sense.

Moreover, "legislative risk," "lack of streamlined tools and processes," and "productivity risk" were the three least critical risks of green retrofit projects. The low RC of "legislative risk" indicates that the respondents did not perceive the change in tax or incentive schemes as a big problem. Indeed, this risk seldom occurred because it received the lowest LO rating. "Lack of streamlined tools and processes" was not viewed as critical because Singapore's government has provided several guides for green construction, such as the "Green and Gracious: Builder Guide" (BCA, 2010c). "Productivity risk" was the least critical risk, indicating that the productivity of green retrofit projects was satisfactory. This was probably because Singapore's government has invested in productivity growth and launched several schemes to raise construction productivity, such as the Productivity Improvement Project (BCA, 2013d).

14.3.2 Risk Criticalities

The respondents were also requested to rate the LO and MI of the 20 risks based on their experience in conventional retrofit projects in Singapore. Thus, the RC values were compared between conventional and green retrofit projects, as shown in Table 14.9.

To check the difference in RC values between conventional and green retrofit projects, the paired *t*-test was adopted. This method is usually used to compare means on the same subject over time or in different circumstances. The analysis results showed that *P*-values of 19 risks were less than .05, indicating significant differences in RC values between conventional and green retrofit projects. The RC values of these 19 risks in green retrofit projects were significantly higher than those in conventional retrofit projects, suggesting that the research hypothesis was supported.

TABLE 14.9 Comparison of RC Between Conventional and Green Retrofit Projects in Singapore

		Conventional		Green		Paired *t*-Test	
Code	Risks	*RC*	*Rank*	*RC*	*Rank*	*Difference*	P-*Value*
R1	Financial risk	9.20	3	16.83	4	7.63	.000*
R2	ROI risk	7.97	11	13.50	11	5.53	.000*
R3	Energy saving uncertainty	6.90	18	11.83	15	4.93	.000*
R4	Accreditation of ESCOs	5.40	20	10.73	16	5.33	.000*
R5	Regulatory risk	10.73	2	17.77	2	7.04	.000*
R6	Legislative risk	7.47	15	10.07	18	2.60	.000*
R7	Warranty risk	7.30	17	10.13	17	2.83	.001*
R8	Market risk	9.10	4	17.00	3	7.90	.000*
R9	Industry risk	8.40	6	13.00	13	4.60	.000*
R10	Delay in project completion	8.17	10	12.17	14	4.00	.000*
R11	Productivity risk	8.33	7	9.43	20	1.10	.065
R12	Manpower supply and availability	7.60	14	13.30	12	5.70	.000*
R13	Material supply and availability	6.03	19	15.30	7	9.27	.000*
R14	Team performance risk	7.40	16	14.40	10	7.00	.000*
R15	Reliability and accuracy of benchmark	9.00	5	14.47	9	5.47	.000*
R16	Quality risk	8.23	8	15.20	8	6.97	.000*

Continued

TABLE 14.9 Comparison of RC Between Conventional and Green Retrofit Projects in Singapore—cont'd

Code	Risks	Conventional		Green		Paired *t*-Test	
		RC	*Rank*	*RC*	*Rank*	*Difference*	*P-Value*
R17	Preretrofit tenants' cooperation risk	7.90	12	15.90	5	8.00	.000*
R18	Postretrofit tenants' cooperation risk	7.77	13	18.97	1	11.20	.000*
R19	Different concerns of stakeholders	11.00	1	15.37	6	4.37	.000*
R20	Lack of streamlined tools and processes	8.23	9	9.93	19	1.70	.008*

Note: * *The paired t-test result is significant at the .05 significance level.*
The Spearman rank correlation coefficient is .408 (P-value=.075).

"Preretrofit tenants' cooperation risk" and "postretrofit tenants' cooperation risk" received low ranks in the conventional retrofit projects, but the differences in RCs between the two groups were great. It is an arduous task to manage tenants to maintain the benefits brought by green retrofits, as a postsurvey interviewee indicated. In conventional retrofit projects, the building owners do not have to demonstrate evidence of energy savings, so the cooperation of tenants is not recognized as crucial.

"Material supply and availability" also received a low risk rank in the conventional group, and the second largest difference in RCs between the two groups. The postsurvey interviewees expressed their struggles to search for green materials and were unaware of the available resources in the market, while common materials are more widely available and known.

The difference in the RCs of "market risk" was the fourth largest. This was possibly because the real estate agents and investors were not clear about the benefits of green building. The buildings with Green Mark Platinum usually experience a noticeable increase in the sale price premiums, compared to those at the Green Mark Certified level (WorldGBC, 2013). The high price premiums tend to make green retrofit buildings unattractive to potential investors (Durmus-Pedini and Ashuri, 2010; Lockwood, 2009).

"Financial risk" received the fifth greatest difference in RCs, although it was ranked high in both groups. Compared with conventional retrofitting, green retrofitting involves investing in energy-saving equipment and technologies, thus having higher upfront costs. Also, the development of an accurate cash flow forecast relies on current knowledge and past experience. Because the Green Mark Scheme has become mandatory since 2008 and green retrofitting has been highlighted since 2009, most practitioners are relatively inexperienced in cost estimation of green retrofit projects.

"Regulatory risk" was ranked second in both groups, but the RC was much lower in the conventional retrofit projects. This was because green retrofit projects should be in compliance with both the basic regulations and those specifically related to the Green Mark Scheme.

"Productivity risk" was the only risk that received a P-value over .05 in the paired t-test, indicating the criticality of this risk was not statistically different between the two types of retrofits.

Furthermore, the Spearman's rank correlation was performed and statistically tested to examine the agreement on the ranking between conventional and green retrofit projects. It is a method of computing a correlation between the ranks of scores between two groups. The correlation is calculated on the ranks of scores, not the scores themselves. As a result, without the consideration of normality or equal variance of data, this method focuses on differences in rank orders of data rather than differences in means. The Spearman's rank correlation coefficient r is calculated by the following formula:

$$r = 1 - \left[6\sum d^2/\left(n^3 - n\right)\right] \qquad (14.1)$$

where d is the difference between ranks assigned to each factor, and n is the number of pairs of ranks. The correlation coefficient of .408 with a P-value of .075 suggested that there was no significant agreement in the risk ranking between the two groups.

14.3.3 Risk Mitigation Measures

In the questionnaire, the respondents were also asked to indicate their endorsement for the 37 risk-mitigation measures that may tackle the risks in green retrofit projects, using a five-point rating scale of 1 = ineffective, 2 = less effective, 3 = somewhat effective, 4 = effective, and 5 = very effective. As Table 14.10 shows, 28 risks obtained significant agreement. The mitigation measures for the most critical risks are discussed below.

In order to obtain tenants' cooperation after retrofits, building owners could increase tenants' awareness of the benefits and provide incentives and rebates to the tenants, such as rental rebates. This measure has been adopted by some commercial buildings in Singapore, such as the 313@Somerest building. This practice is also consistent with the concept of the green lease. In addition, as for the mitigation measures for the preretrofit tenants' cooperation risk, the respondents did not believe it was feasible to perform work outside standard working hours and arrange relocation for tenants because these two measures would increase the project cost and the tenants might still be reluctant to cooperate. Thus, it is desirable to first get the tenants to understand the rationale behind the retrofit works.

To mitigate the regulatory risk, it is reasonable to hire consultants for advice, who have the experience and knowledge in managing green retrofit projects. These consultants should be registered under the Green Mark Professional Scheme.

As for the mitigation measures for market risk, building owners could make the public aware of the possible benefits of green buildings through media and use specific evidence to demonstrate those benefits. Previous studies have emphasized educating the public through media (Durmus-Pedini and Ashuri, 2010; WorldGBC, 2013). However, collecting the evidence of building performance and energy savings received a higher score from the respondents, indicating that practitioners preferred real case studies to just education.

To handle the financial risk, practitioners could choose to retrofit in phases. By performing a green retrofit project in phases, the size and first cost would be noticeably lower, and the impact on companies' cash flow would not be as great. Also, Singapore's government has been very supportive of green retrofitting and has provided several financial incentive schemes. Hence, these incentive schemes could be utilized to mitigate financial risk. These two measures received high scores from the respondents, indicating their applicability.

To integrate various stakeholders, Design & Build (DB), as an integration delivery method, is a viable solution to improve communication and minimize

TABLE 14.10 Risk Mitigation Measures in Green Retrofit Projects

Risk	Risk Mitigation Measures		Mean	*P*-Value
R01	M1a	Retrofit in stages	4.23	.000*
	M1b	Share the risk with other parties by contractual agreement	3.67	.000*
	M1c	Tap on government and financial incentives (e.g., GMIS-EB, GMIS-EBP, and BREEF)	4.03	.000*
R02	M2	Document building performance and savings to establish a proof of higher performance	3.60	.000*
R03	M3	Establish a performance based contract with the consultants	3.97	.000*
R04	M4a	Check ESCOs' credibility with reference to the list of ESCOs provided by the government	4.07	.000*
	M4b	Seek background information on the type of past projects handled by the ESCOs	3.50	.001*
R05	M5	Hire consultants for advice	3.73	.000*
R06	M6a	Obtain legal advice (e.g., when drafting contract)	3.80	.000*
	M6b	Clearly define the liability	3.63	.000*
R07	M7	Transfer risk to a third party (e.g., purchase insurance)	3.83	.000*
R08	M8a	Educate the public on green building through use of media	3.53	.000*
	M8b	Document performance and savings earned as benefit evidence	3.77	.000*
R09	M9a	Educate professionals' knowledge of green building	3.27	.030*
	M9b	Government to sponsor more green building courses for professionals	3.30	.163

Continued

TABLE 14.10 Risk Mitigation Measures in Green Retrofit Projects—cont'd

Risk	Risk Mitigation Measures		Mean	*P*-Value
R10	M10a	Adopt technologies (e.g., BIM) to manage projects	3.00	1.000
	M10b	Set attainable milestones and monitor them regularly. If necessary, expedite the project	3.80	.000*
	M10c	Understand and set reasonable buffer time	3.73	.000*
R11	M11	Implement good resource scheduling and planning to maximize productivity	3.80	.000*
R12	M12a	Increase the usage of automation and machinery	2.90	.687
	M12b	Change Singaporeans mindset and encourage them to take up the job	2.60	.083
R13	M13a	Procure materials which are manufactured and available in either Singapore or the neighboring countries	2.60	.016*
	M13b	Carefully schedule material supplies	3.73	.000*
	M13c	Industry institutions and government to provide a portal or list of green materials available	3.93	.000*
	M13d	Obtain government support and invest in green material and technology research and development	3.80	.000*
R14	M14a	Improve team members' communication and integration	3.43	.000*
	M14b	Select the Design & Build delivery method	3.13	.403
R15	M15	Prepare a "knowledge portal" for historical data and lesson learnt to be recorded	3.70	.000*
R16	M16a	Use tested and certified green materials	3.50	.009*
	M16b	Conduct routine checks to ensure the desired level of workmanship are met	3.37	.062

TABLE 14.10 Risk Mitigation Measures in Green Retrofit Projects—cont'd

Risk	Risk Mitigation Measures		Mean	*P*-Value
R17	M17a	Perform retrofit work outside standard working hours	2.83	.283
	M17b	Arrange relocation for the tenants	2.97	.787
	M17c	Explain the rationale for retrofitting existing building to green building	3.57	.000*
R18	M18a	Increase awareness of the benefits of retrofitting existing building to green building	3.50	.001*
	M18b	Provide incentives and rebates to the tenants (e.g., rental rebate)	3.97	.000*
R19	M19	Engage in the Design & Build delivery method	3.27	.043*
R20	M20	Government to sponsor more green building research	3.33	.030*

*Note: * The one-sample* t*-test result is significant at the .05 significance level.*

the risk of alienating any stakeholder (Kibert, 2012; Glavinich, 2008). This is because DB allows early discipline integration (Xia and Chan, 2011; Xia et al., 2011), and thus various stakeholders' concerns can be put across.

Moreover, to ensure material supply and availability, a portal or list of green materials available can be helpful because postsurvey interviewees revealed difficulties in searching for suitable green materials available in the market. A careful schedule of material supply is also reasonable to handle this risk (Hwang et al., 2015b). Another approach, with agreement from the respondents, is to develop green materials through research projects funded by the government, thus reducing Singapore's dependence on imports of green materials.

Furthermore, to ensure the quality of green retrofit projects, tested and certified green materials should be used. However, it was not feasible to conduct routine checks, as revealed by its nonsignificant score. This was not surprising as routine checks would not be able to reduce "green washing."

14.4 ANALYSIS RESULTS AND DISCUSSIONS: GREEN RESIDENTIAL CONSTRUCTION PROJECTS

14.4.1 Risk Ranking

Prior to analyzing the data, the Cronbach's alpha was calculated to check the internal consistency or reliability of the data. Following the instruction of Nunnally et al. (1967), the alpha should be at least 0.7 for a scale to be reliable. For this survey, the alpha values of LO and MI in green residential building projects were 0.956 and 0.957, respectively, indicating that the data collected from the questionnaire survey were of sufficient reliability.

Using Eqs. (3.1), (3.2), the LO, MI, and RC values of each risk were calculated and presented in Table 14.11. Regarding the LO values, 28 risks were assessed above 3, suggesting they were fairly likely to occur in green residential construction projects. In particular, the top five risks in LO values were: "complex procedures to obtain approvals," "overlooked initial high cost," "unclear requirements," "employment and material constraints," and "technical issues." As for the MI values, 37 risks received MI values above 3, suggesting they had fairly significant impacts on green residential construction projects. In particular, the top five most impactful risks were: "complex procedures to obtain approvals," "overlooked initial high cost," "unfamiliarity of job requirement," "lack of availability causing delay," and "shortage of funds." In terms of RC values, 35 risks scored above 9, suggesting they were critical risks in green residential construction projects. In particular, the top five critical risks were: "complex procedures to obtain approvals," "overlooked initial high cost," "unclear requirements of owners," "employment constraint," and "lack of availability of green materials and equipment," and all their RC values were greater than 14, indicating that they were extremely critical to green residential construction projects.

TABLE 14.11 LO, MI, and RC of Risks in Green Residential Projects in Singapore

	LO		MI		RC	
Code	*Mean*	*Rank*	*Mean*	*Rank*	*Mean*	*Rank*
R1	1.87	41	3.23	26	6.10	40
R2	2.23	40	3.17	32	7.47	39
R3	2.93	30	3.13	35	9.80	33
R4	3.10	27	3.23	26	10.97	26
R5	1.50	42	1.90	42	3.37	41
R6	4.20	1	4.00	1	17.27	1
R7	3.20	20	3.40	14	11.30	22
R8	3.83	4	3.73	6	14.47	4
R9	3.17	23	3.20	30	10.63	27
R10	2.60	38	3.30	22	8.93	36
R11	2.93	30	3.40	14	10.27	29
R12	3.13	25	3.27	25	10.60	28
R13	2.40	39	2.77	41	7.60	38
R14	2.97	29	3.77	5	11.50	18
R15	3.90	3	3.73	6	15.07	3
R16	2.77	35	2.87	40	8.63	37
R17	2.77	35	2.97	38	9.13	35
R18	2.77	35	2.90	39	8.93	36
R19	3.53	8	3.43	12	12.87	9
R20	3.13	25	3.23	26	11.53	17
R21	2.90	32	3.33	21	9.87	31
R22	3.53	8	3.70	8	13.53	6
R23	2.87	33	3.40	14	9.73	34
R24	3.37	14	3.87	3	13.50	7
R25	3.03	28	3.60	9	11.43	19
R26	4.13	2	3.97	2	16.67	2
R27	3.70	5	3.40	14	13.00	8
R28	3.57	6	3.80	4	14.17	5

Continued

TABLE 14.11 LO, MI, and RC of Risks in Green Residential Projects in Singapore—cont'd

	LO		MI		RC	
Code	*Mean*	*Rank*	*Mean*	*Rank*	*Mean*	*Rank*
R29	3.33	16	3.37	20	11.67	15
R30	3.53	8	3.00	37	11.23	23
R31	3.23	19	3.20	30	11.33	21
R32	3.20	20	3.17	32	11.00	25
R33	3.47	11	3.30	22	12.23	12
R34	3.27	18	3.43	12	11.37	20
R35	3.17	23	3.10	36	10.03	30
R36	3.20	20	3.53	10	11.83	14
R37	2.80	34	3.17	32	9.83	32
R38	3.30	17	3.23	26	11.17	24
R39	3.37	14	3.30	22	11.57	16
R40	3.57	6	3.40	14	12.77	10
R41	3.40	12	3.50	11	12.63	11
R42	3.40	12	3.40	14	12.00	13

"Complex procedures to obtain approvals" occupied the top position in the risk ranking with an RC value of 17.27, owing to its highest LO and MI values. This was perhaps because green residential construction projects tend to involve innovative and complicated construction methods, which result in more lengthy planning approval and permit procedures (Zhao et al., 2016). Shiers et al. (2006) also identified this risk in green construction projects and labeled it an "extra step" that wasted considerable time and caused serious disturbances. Moreover, a long processing time for obtaining approval compromises the on-time delivery of green residential construction projects and thus increases the likelihood of liquidated damages and poor product quality (Zou and Couani, 2012). This risk has also been assessed as the most critical risk in Chinese green construction projects by Qin et al. (2016).

"Overlooked initial high cost" was ranked second with an RC value of 16.67, attributed to its second highest LO and MI values. Compared to traditional residential construction projects, green residential construction projects frequently involve a high initial cost that can be attributed to the enormous upfront costs of green technologies and materials and the additional consultancy services with green constructions (Zou and Couani, 2012; Hwang and Tan, 2012). However, this issue has been often ignored by industry practitioners in Singapore when they estimate the project cost. This is probably because the green construction industry is still in its early stages in Singapore and many local industry practitioners haven't yet gained enough experience in addressing green construction. This result echoed the findings from Robichaud and Anantatmula (2010) who also reached a similar conclusion.

"Unclear requirements of owners" received the third position with an RC value of 15.07. This risk received the third highest LO, suggesting that it had a high chance of occurring. During the development of green residential construction projects, the complexity involved might make the owners unable to give clear and specific requirements to designers and contractors (Zhao et al., 2016), which would inevitably increase the risks in the projects. For instance, unclear requirements from the clients might result in misinterpretations on the part of the designers or misunderstanding of the clients' real intentions, leading to numerous design changes and reworks. This result also echoed Hwang et al. (2015d) who recognized "unclear requirements of owners" as a critical issue affecting the schedule performance of green construction projects in Singapore.

"Employment constraint" was ranked fourth with an RC value of 14.47. Being a country with limited human resources, Singapore is always leveraging foreign workforces to ensure its economic growth (Low, 2002). Particularly, the construction industry of Singapore is heavily reliant on foreign manpower as local workers are hardly interested in this laborious industry (Ong, 2014). Nonetheless, the Singapore government has established some control mechanisms to regulate the number of foreign workers so as to avoid the local Singaporeans being priced out of the job market. Unfortunately, foreign workers with green residential construction experience typically belong to the unskilled or semi-skilled workforce, which are under the strict control of the Minister of Manpower (Ong, 2014). Thus, the local green residential construction projects might face an issue of lack of sufficient manpower owing to this employment constraint.

"Lack of availability of green materials and equipment" was assessed as the fifth most critical risk with an RC value of 14.17. It has been widely recognized in the literature that adequate material and equipment supply is an important determinant of successful construction projects, and a shortage of material and equipment would definitely jeopardize the success of projects (Ogunlana et al., 1996). In Singapore, the construction equipment, materials, and even plants designated for the green residential buildings, need to be imported from overseas, which normally requires a long period to be delivered on site (Hwang and Leong, 2013; Zhao et al., 2016). Therefore, any issue related to the delivery of imported equipment and materials, especially for one concerning the critical activities in the project schedule, would significantly affect the successful delivery of the project. This result is in line with the findings from Zhao et al. (2016), who also emphasized that "availability of materials and equipment" was a significant risk requiring additional attention in green construction projects.

14.4.2 Risk Criticalities

The respondents were also requested to rate the LO and MI of the 42 risks in the context of conventional residential construction projects in Singapore, based on their practice experience. Thus, the RC values of the 42 risks in conventional residential construction projects were also calculated and then compared with those in green residential construction projects, as shown in Table 14.12.

To look into the possible difference in RC values between conventional and green residential construction projects, the Wilcoxon signed-rank test was used. This method is a nonparametric statistical test to compare two sets of scores that come from the same participants, without requiring the data to be normally distributed. This test can also be used to investigate the possible changes in scores

TABLE 14.12 Comparison of RC Between Conventional and Green Residential Projects in Singapore

	Traditional		Green		Wilcoxon Signed-Rank Test	
Code	*RC*	*Rank*	*RC*	*Rank*	*Difference*	P-*Value*
R1	5.47	33	6.10	40	0.63	.059
R2	6.00	30	7.47	39	1.47	.137
R3	7.87	18	9.80	33	1.93	.003*
R4	7.70	19	10.97	26	3.27	.034*
R5	3.03	36	3.37	41	0.34	.776
R6	9.73	2	17.27	1	7.54	.000*
R7	9.27	5	11.30	22	2.03	.030*
R8	10.00	1	14.47	4	4.47	.000*
R9	8.53	11	10.63	27	2.1	.051
R10	6.60	26	8.93	36	2.33	.003*
R11	7.93	17	10.27	29	2.34	.003*
R12	8.57	10	10.60	28	2.03	.000*
R13	7.27	24	7.60	38	0.33	.916
R14	8.33	14	11.50	18	3.17	.000*
R15	8.40	13	15.07	3	6.67	.000*

TABLE 14.12 Comparison of RC Between Conventional and Green Residential Projects in Singapore—cont'd

	Traditional		Green		Wilcoxon Signed-Rank Test	
Code	*RC*	*Rank*	*RC*	*Rank*	*Difference*	P-*Value*
R16	6.20	28	8.63	37	2.43	.015*
R17	7.67	20	9.13	35	1.46	.002*
R18	5.37	34	8.93	36	3.56	.000*
R19	7.63	21	12.87	9	5.24	.000*
R20	4.70	35	11.53	17	6.83	.000*
R21	9.10	7	9.87	31	0.77	.325
R22	5.50	32	13.53	6	8.03	.000*
R23	6.60	26	9.73	34	3.13	.001*
R24	7.10	25	13.50	7	6.4	.000*
R25	6.03	29	11.43	19	5.4	.000*
R26	8.60	9	16.67	2	8.07	.000*
R27	7.53	22	13.00	8	5.47	.000*
R28	7.40	23	14.17	5	6.77	.000*
R29	6.43	27	11.67	15	5.24	.000*
R30	5.87	31	11.23	23	5.36	.000*
R31	6.03	29	11.33	21	5.3	.000*
R32	8.50	12	11.00	25	2.5	.011*
R33	9.27	5	12.23	12	2.96	.004*
R34	8.27	15	11.37	20	3.1	.002*
R35	9.00	8	10.03	30	1.03	.168
R36	9.30	4	11.83	14	2.53	.021*
R37	7.53	22	9.83	32	2.3	.011*
R38	8.40	13	11.17	24	2.77	.013*
R39	9.13	6	11.57	16	2.44	.001*
R40	8.00	16	12.77	10	4.77	.000*
R41	9.57	3	12.63	11	3.06	.000*
R42	6.20	28	12.00	13	5.8	.000*

Note: * *The Wilcoxon signed-rank test is significant at the .05 significance level.*

from one time point to another, or when participants are subjected to more than one condition (Ameyaw et al., 2016). For this survey, the test results reported that the P-values of 35 risks were less than .05, suggesting significant differences in RC values among the majority of risks between green and conventional residential construction projects. This result means that the hypothesis of this chapter is supported. Moreover, the RC values of these 35 risks in green residential construction projects were statistically greater than those in conventional residential construction projects, implying that green residential construction projects face risks at a greater magnitude than conventional residential construction projects.

"Overlooked initial high cost" received the greatest difference between the two types of projects. The RC value of this risk in green residential construction projects (16.67) is almost two times of that in conventional residential construction projects (8.60). This was probably because, compared to conventional residential construction projects, green residential construction projects require considerable upfront expenditures on green technologies, materials, and equipment (Hwang and Tan, 2012), which eventually would take a toll on the construction budget. This result is also supported by Zou and Couani (2012) who claimed that "perceived higher upfront costs" was the largest obstacle in green construction development.

"Lack of qualified professionals with proper design expertise" received a low rank in conventional residential construction projects (32), but a high rank in green residential construction projects (6). This result implies that the lack of qualified professionals in green construction design is a very critical risk to green residential construction projects. Green residential construction projects require skilled design professionals to handle the specialized green and sustainable designs. The reality in Singapore is that competent and experienced local green design professionals are extremely small in number (Hwang et al., 2015a). Moreover, the Ministry of Manpower of Singapore has recently increased the foreign worker levies and cut the quotas of foreign workers in the construction industry (Nair, 2016), which has aggravated the unavailability of competent manpower.

"Complex procedures to obtain approvals" obtained the third greatest difference in RCs. Although this risk received high ranks in both groups, its RC values were significantly different between two types of projects: 17.27 with green versus 9.73 with conventional. Because green residential construction projects involve more innovative technologies, materials, and equipment compared to conventional construction projects, they have to undergo a stricter approval process imposed by the construction authority (Zou and Couani, 2012; Zhao et al., 2016), which inevitably makes the processing time longer than that for conventional residential construction projects.

"Being fined for failing to achieve Green Mark standards" received the fourth greatest difference in the RC assessment, and in particular, its RC value in green projects (11.53) was much higher than that in conventional projects

(4.70). This was probably due to Singapore's Construction Act, which has required that any new constructions, and existing ones that undergo major retrofitting, must achieve the minimum Green Mark Certified Level; otherwise a certain amount of fines will be imposed (BCA, 2015f). For instance, as enforced by the BCA, construction owners can be subject to a fine of up to SGD 100,000 if they fail to achieve the minimum Green Mark standard in their installation of a new cooling system (Ismail, 2013).

"Lack of availability of green materials and equipment" was ranked fifth in the RC difference, and it received a low rank (23rd) in conventional projects, but a high rank (5th) in green projects. This was probably because the green construction industry in Singapore is a young industry and thus the supply of green materials and equipment might be still limited; by contrast, the conventional construction industry is already fully mature and thus the common materials and equipment are more widely available (Hwang et al., 2015a). This result is also in line with Hwang and Leong (2013), who argued that "material supply and availability" is more critical in green construction projects compared to conventional construction projects.

14.4.3 Risk Mitigation Measures

The effectiveness of the 14 risk mitigation measures generated from literature were also evaluated. As Table 14.13 shows, all the 14 risk mitigation measures received mean values higher than 3, indicating all of them were effective in tackling risks in green residential construction projects. Those mitigation measures that received evaluations above 4 are discussed below.

"Improve communication and coordination among contracting parties" was assessed as the most effective measure with the highest evaluation of 4.57. To ensure the success of a green residential construction project, a higher level of communication is demanded among the contracting parties as compared with conventional residential construction projects. This is because the green residential construction projects normally require a multidisciplinary team with a more comprehensive professional composition to handle the complicated and innovative technologies, equipment, and materials adopted in such projects (Yang and Zou, 2014). Any information isolation among the team members will probably raise various issues, such as rework, delay, and cost overrun. This result is also in line with Hwang and Tan (2012), who stated that "project team's communication" is an effective solution to overcome the obstacles in green construction projects.

"Understand owner's goal of the Green Mark Standard" received the second highest value (4.27) in the effectiveness evaluation. In Singapore, each new residential construction is mandated to achieve some Green Mark Standard (BCA, 2015f). Thus, it is crucial for the designer, consultant, and contractor to understand the owner's goal of Green Mark Standard for the construction before they start working on the project. Any negligence or misunderstanding of the

TABLE 14.13 Risk Mitigation Measures in Green Residential Construction Projects

Code	Risk Mitigation Measure	Mean	Rank
S1	Front end planning	3.83	6
S2	Constant design evaluation and verifications	3.53	9
S3	Implement passive design instead of complicated active building design	3.50	10
S4	Work with experienced insurance agent to receive better coverage protection	3.30	12
S5	Allow for contingency funds	4.13	4
S6	Effective change management	3.10	13
S7	Enhanced communication tool for better collaboration (e.g., BIM software)	3.4	11
S8	Contract language to be precise and give provision to limit each parties' liabilities	3.67	8
S9	Understand owner's goal of the Green Mark Standard	4.27	2
S10	Devote adequate resources to planning and research	3.90	5
S11	Improve communication and coordination among contracting parties	4.57	1
S12	Use past successful green residential projects as references	4.20	3
S13	Develop training programs to upgrade workers' skills and knowledge on new technologies and materials	4.20	3
S14	Communicate about targeted green mark rating and ways to achieve that with a clear roles and responsibilities chart	3.77	7

owner's Green Mark goal can cause the project to fail to achieve the due green certification and cause considerable disputes, claims and liquidated damages (Qin et al., 2016; Tollin, 2011). Zou and Couani (2012) obtained a similar conclusion that communicating green construction objectives clearly to all the project team members is vital to secure the success of a green construction project.

"Use past successful green residential projects as references" was assessed as the third most effective measure with an evaluation of 4.20. Referring to the successful experiences of past projects is an effective measure to mitigate risks in new construction projects (Pinto and Slevin, 1987; Cooke-Davies, 2002). So far, Singapore has accumulated some valuable experience in developing green residential constructions since it introduced the Green Mark for Residential Constructions in 2011 (BCA, 2016g). These experiences can render considerable help to new green residential construction projects in mitigating risks and achieving better project performance. This result was comparable to Zou and Couani (2012), who stated that experience accumulation was an important strategy to reduce risks in the green construction supply chain.

"Develop training programs to upgrade workers' skill and knowledge on new technologies and materials" was also assessed as the third most effective measure with a value of 4.20. During the development of green construction projects, one major obstacle is that the practitioners might be unfamiliar with the innovative technologies and materials adopted in such projects (Häkkinen and Belloni, 2011; Hwang and Tan, 2012). The practitioners' improper operations might cause various issues to projects, such as equipment failure and the waste of valuable green materials. Thus, it is vital to develop a series of training programs to help practitioners gain more skills and knowledge of green construction. Currently, the BCA has rolled out a number of green courses (e.g., Green Mark Professional course and Green Mark Facilities Professional course) to help the local industry advance their knowledge and capability in undertaking green construction projects, which have become very popular with local construction companies (BCA, 2015d).

"Allow for contingency funds" received the fourth highest value (4.13) in the effectiveness evaluation. This risk mitigation measure was highlighted because the complex nature of green residential construction projects makes an exact project budget impossible to forecast accurately. In addition, the innovative and complicated green technologies adopted in green residential construction projects might require additional testing and inspection during the construction process (Häkkinen and Belloni, 2011), which would also lead to additional cost beyond the original project estimate. Therefore, it is extremely necessary to set aside some contingency funds to mitigate some unexpected but possible risks. In fact, contingency funds have also been used widely by traditional construction projects to manage their risks (Akintoye and MacLeod, 1997; Ford, 2002; Thal Jr. et al., 2010).

14.5 ANALYSIS RESULTS AND DISCUSSIONS: GREEN COMMERCIAL CONSTRUCTION PROJECTS

14.5.1 Risk Ranking

The LO, MI, and RC values of each risk in green commercial construction projects in Singapore are calculated using Eqs. (3.1), (3.2). The calculated results are shown in Table 14.14. The widely-used Cronbach's alpha was calculated to check data reliability, with the aid of SPSS Statistics 17.0. The alpha values of

TABLE 14.14 LO, MI, and RC of Risks in Green Commercial Construction Projects

	LO			MI			RC	
Code	*Mean*	*P-Value*	*Rank*	*Mean*	*P-Value*	*Rank*	*Mean*	*Rank*
A1	3.15	.043*	16	3.89	.000*	19	12.22	17
A2	2.33	.000*	32	4.48	.000*	1	10.63	29
A3	3.00	.000	23	4.04	.000*	12	12.41	15
A4	2.44	.041*	31	4.30	.000*	4	10.04	31
A5	2.89	.083	27	4.22	.000*	6	12.33	16
A6	3.15	.527	16	4.44	.000*	2	14.11	5
A7	3.56	.000*	3	4.15	.000*	9	14.96	3
B1	3.22	.056	13	4.15	.000*	9	13.22	12
B2	3.56	.000*	3	3.96	.000*	17	14.56	4
B3	3.41	.001*	8	3.96	.000*	17	13.56	8
B4	3.48	.009*	6	4.00	.000*	15	13.78	6
C1	1.70	.000*	34	3.37	.284	29	6.19	34
C2	3.19	.379	15	4.04	.000*	12	13.19	13
C3	2.48	.041*	30	4.00	.000*	15	9.26	32
C4	3.22	.282	13	4.33	.000*	3	13.56	8
C5	1.85	.001*	33	3.89	.000*	19	7.00	33
C6	3.11	.600	18	4.30	.000*	4	13.30	11

C7	2.85	.515	28	3.11	.746	34	10.30	30
C8	3.26	.316	10	3.30	.266	33	11.78	23
C9	3.56	.001*	3	3.85	.000*	22	13.52	10
D1	3.78	.000*	1	4.22	.000*	6	16.30	1
D2	3.00	1.000	23	4.04	.000*	12	11.89	21
D3	2.85	.294	28	4.11	.000*	11	12.22	17
D4	3.74	.000*	2	4.19	.000*	8	15.96	2
D5	3.07	.537	20	3.89	.000*	19	12.00	19
D6	3.26	.166	10	3.70	.000*	25	12.00	19
E1	3.26	.148	10	3.85	.000*	22	12.74	14
E2	3.44	.003*	7	3.81	.000*	24	13.67	7
E3	2.93	.746	26	3.37	.152	29	10.96	27
E4	3.00	1.000	23	3.44	.103	28	10.93	28
F1	3.37	.015*	9	3.37	.002*	29	11.52	24
F2	3.04	.852	22	3.56	.037*	26	11.85	22
F3	3.07	.703	20	3.37	.152	29	11.19	26
G1	1.63	.000*	35	2.74	.463	35	5.81	35
G2	3.11	.376	18	3.52	.055	27	11.33	25

*Note: * The one sample t-test result is significant at the significance level of .05 (two-tailed).*

LO and MI in green commercial construction projects were 0.957 and 0.987, respectively, both of which are considerably greater than the threshold of 0.7, proving that the survey data are reliable (Nunnally et al., 1967). A one sample *t*-test was subsequently carried out to verify whether each risk factor occurs significantly in, or has a significant impact on, green commercial construction projects. The one sample *t*-test is usually used to examine the mean difference between the sample and the known value of the population mean (Lam et al., 2011). For this chapter, the test value was established as 3, the median value of the Five-point Likert scale. Test results showed that the majority of risks occur significantly in green commercial construction projects, except for A2, A4, C1, C3, C5, and G1; and that all the risks have significant impacts on green commercial construction projects, considering that their LO and MI values are statistically greater or equal to the test value of 3 at a 95% confidence interval (Zhao et al., 2013b).

According to the LO and MI prioritization, the top five risk factors in the likelihood of occurrence in green commercial construction projects are D1 "inflation," D4 "currency and interest rate volatility worsened by the import of green materials," A7 "durability of green materials," B2 "damages caused by human error," and C9 "delay caused by the frequent meetings with green specialists." Meanwhile, the top five risk factors that have the most significant impact on green commercial construction projects are A2 "insufficient or incorrect design information," A6 "material suitability and accessibility and shortage of green material," C4 "customer expectation management," A4 "poor definition of scope and changes in scope," and C6 "limited availability and reliability of green suppliers." As for RC values, the top five critical risk factors in green commercial construction projects were found to be D1 "inflation," D4 "currency and interest rate volatility worsened by the import of green materials," A7 "durability of green materials," B2 "damages caused by human error," and A6 "material suitability and accessibility and shortage of green material." All these factors will be discussed in detail as follows.

Two financial-related risk factors, namely, D1 "inflation" and D4 "currency and interest rate volatility worsened by the import of green materials," have been assessed as the top two critical risk factors in green commercial construction projects in Singapore. Inflation is the increase in the price level of commodities and the decrease of the purchasing power of currencies, which can eventually cause cost overrun to projects (Chan et al., 2011a). Although the inflation rate in Singapore has been kept at below zero in the last 2 years, it used to be up to 6% between 2011 and 2013 (Trading Economics, 2016). Thus, subconsciously, the respondents still believe inflation to be a critical risk factor in building projects. Singapore is also a small country with limited natural resources (Hawksford, 2009). The country has to import green materials and equipment to satisfy its domestic construction of green commercial projects. Therefore, the volatility of currency and of interest rates in the country will incur an increase in the price of imported materials and equipment and probably cause a project cost overrun. To relieve the negative impact of these two financial risk factors, respondents claimed that the involvement of extension quotations in procurement strategy and a timely updated cost contingency plan would be helpful.

A7 "durability of green materials" was assessed as the third critical risk factor in green commercial construction projects. This ranking may be attributed to the tendency of such projects always to use innovative materials that are newly developed to achieve improved performance of sustainability (Reijnders and Van Roekel, 1999). However, such materials may lack sufficient testing and this may lead durability issues (Häkkinen and Belloni, 2011), which may eventually undermine the successful delivery of green commercial construction projects. A similar concern was also expressed by Lam et al. (2009) and Shi et al. (2013) that green material durability has become a significant barrier that prevents the construction industry from implementing green construction strategies. To reduce the risk in this regard, respondents suggested carrying out in-depth research on new materials and technologies that are going to be adopted in the project to guarantee their effectiveness.

B2 "damages caused by human error" was ranked fourth in the criticality assessment in green commercial construction projects. Such risk is common and unavoidable in any type of project (Wibowo and Kochendoerfer, 2011). However, the risk is even more critical to green commercial construction projects, because technologies and equipment adopted in these projects always get updated quickly and may be beyond the common knowledge of frontline workers (Kang et al., 2013). Therefore, damages may arise due to the workers' inadvertent errors when dealing with new equipment or instruments. To lower the likelihood of this risk factor, respondents suggested implementing more training and education. Penalties were also suggested to deter noncompliance with operating regulations wherever necessary.

A6 "material suitability and accessibility and shortage of green material" received the fifth ranking in criticality assessment. Respondents expressed their concerns about this critical risk factor in two respects, namely, the absence of green materials and the poor supply of green materials. The former may occur due to the development lag of green materials with green technologies, while the latter may be caused by a poor supply management of green materials, which might be caused by poor forecasting, planning, low inventory levels, poor communication of information, or dependence on a single supplier. To solve the former, a close monitoring of the latest developments in green materials was recommended by the respondents. For the latter, maintaining a close and strong client-supplier relationship, as well as full monitoring of green material production, delivery, and fabrication, were suggested. These findings were consistent with the arguments of Shi et al. (2013) that uncertain green material supplies is a significant barrier against green construction.

14.5.2 Risk Criticalities

Respondents were also requested to rate the LO and MI of the 35 risk factors in terms of their experience in conventional commercial construction projects in Singapore. Thus, RC values were compared between green and conventional commercial construction projects, as indicated in Table 14.15.

TABLE 14.15 Comparison of RC Between Conventional and Green Commercial Construction Projects

	Green		Conventional		Paired *t*-Test	
Code	*RC*	*Rank*	*RC*	*Rank*	*Difference*	P-*Value*
A1	12.22	17	15.63	3	3.41	.001*
A2	10.63	29	11.48	14	0.85	.436
A3	12.41	15	17.59	1	5.18	.000*
A4	10.04	31	10.52	18	0.48	.667
A5	12.33	16	12.52	10	0.19	.863
A6	14.11	5	8.63	26	5.48	.001*
A7	14.96	3	10.07	19	4.89	.000*
B1	13.22	12	13.74	7	0.52	.403
B2	14.56	4	15.15	4	0.59	.355
B3	13.56	8	10.07	19	3.49	.001*
B4	13.78	6	8.81	25	4.97	.001*
C1	6.19	34	5.89	34	0.3	.641
C2	13.19	13	12.44	11	0.75	.262
C3	9.26	32	9.81	23	0.55	.322
C4	13.56	8	14.00	6	0.44	.416
C5	7.00	33	4.78	35	2.22	.106
C6	13.30	11	9.89	22	3.41	.036*
C7	10.30	30	6.63	32	3.67	.013*
C8	11.78	23	6.81	31	4.97	.001*
C9	13.52	10	7.74	28	5.78	.000*
D1	16.30	1	17.48	2	1.18	.200
D2	11.89	21	10.07	19	1.82	.023*
D3	12.22	17	13.00	8	0.78	.420

TABLE 14.15 Comparison of RC Between Conventional and Green Commercial Construction Projects—cont'd

	Green		Conventional		Paired *t*-Test	
Code	*RC*	*Rank*	*RC*	*Rank*	*Difference*	P-*Value*
D4	15.96	2	11.44	15	4.52	.004*
D5	12.00	19	7.56	29	4.44	.004*
D6	12.00	19	6.33	33	5.67	.001*
E1	12.74	14	14.63	5	1.89	.114
E2	13.67	7	11.85	13	1.82	.270
E3	10.96	27	8.52	27	2.44	.092
E4	10.93	28	9.59	24	1.34	.404
F1	11.52	24	12.26	12	0.74	.170
F2	11.85	22	12.93	9	1.08	.244
F3	11.19	26	11.44	15	0.25	.795
G1	5.81	35	7.15	30	1.34	.201
G2	11.33	25	11.15	17	0.18	.873

*Note: * The paired sample* t-test *result is significant at the significance level of .05 (two-tailed).*

To check the difference in RC values between green and conventional commercial construction projects, the paired sample *t*-test was adopted. This method is usually used to compare means on the same subject over time or in different circumstances (Hwang et al., 2015d; Shi et al., 2013). The analysis results showed that the *P*-values of 14 risks were less than .05, indicating significant differences in RC values between green and conventional commercial construction projects. Specifically, A1 and A3, which respectively refer to design change and poor construction quality, scored statistically higher in conventional than in green commercial construction projects, indicating they are more critical to the former. This finding is probably attributed to the two risk factors being extremely common in conventional commercial construction projects. Ireland (1985) also emphasized that design change and poor quality performance are two primary causes of failure in commercial construction projects.

The 12 remaining risk factors, namely, A6, A7, B3, B4, C6, C7, C8, C9, D2, D4, D5, and D6, all scored statistically higher in green commercial construction projects than in conventional ones, indicating they are more critical to the former. These risk factors mainly refer to the issues of green material shortage and durability, poor comprehension of green specifications in contract details, resistance from stakeholders to adopt green ideas, complexity originating from green construction technologies, and poor cost management due to the use of green material and equipment. This result is consistent with previous survey results in Hong Kong, China, and Singapore that green construction projects face more challenges and risks than traditional ones regarding green construction ideas, materials, technologies, and costs (Ofori and Kien, 2004; Shi et al., 2013; Hwang and Ng, 2013; Hwang and Tan, 2012; Lam et al., 2009). In addition, the obvious contrast in the number of risk factors between the two different groups, namely, 2 versus 12, revealed that green commercial construction projects have raised more concern from respondents.

14.5.3 Risk Mitigation Measures

The last section of the questionnaire survey requested respondents to list the mitigation measures that their companies have taken to tackle risk factors in green commercial construction projects. According to the respondent feedback, all the companies have established and undertaken such mitigation measures. The mitigation measures that were mentioned most by respondents in different risk categories are discussed in this section. Table 14.16 presents these mitigation measures, the descriptions of measures, and the frequency results.

Within the technical risk, two mitigation measures were tied in highest frequency. Specifically, "inspecting quality frequently with a detailed checklist" was suggested to address the risk factor A3 "improper quality control and defective work." Respondents suggested that frequent on-site inspections are useful to ensure construction quality and that a detailed checklist of work quality could be devised and utilized when conducting on-site inspections. Another mitigation measure, namely, "ensuring sound project design and early briefing," was recommended to address the risk factor A4 "poor definition of scope and changes in scope." Respondents believed that by taking this measure, mistakes and discrepancies in the design could be spotted earlier and corrections performed promptly; thus, changes in the construction stage could be reduced significantly.

The mitigation measure mentioned with the highest frequency in the labor risk is "guaranteeing a rigorous contractor selection," which is meant to address the risk factor B3 "shortage of skilled and experienced professionals." This intent of this measure is to conduct rigorous contractor selection when awarding a contract and ensure that the awardee has sufficient skilled professionals as well as a good track record. This measure was also strengthened by Hwang et al. (2013a) who asserted that a comprehensive analysis of contractors' capability should be done before awarding the contract.

The mitigation measure mentioned most within the management risk is "managing customer expectations soundly," which targets the risk factor C4 "customer expectation management." The ultimate customers of a commercial

TABLE 14.16 Risk Mitigation Measures in Green Commercial Projects

No.	Mitigation Measure	Description	Targeted Risk Factor	Frequency	Percentage
1	Inspecting quality frequently with a detailed checklist	Devise a detailed checklist for the on-site inspection and engage the inspection more frequently	A3	11	40.7
2	Ensuring sound project design and early briefing	Ensure a sound design for the project and conduct a full briefing among stakeholders in the early stages	A4	11	40.7
3	Guaranteeing a rigorous contractor selection	Rigorous selection of contractor with good track record	B3	8	29.6
4	Managing customer expectations soundly	Manage the customer expectation with consultants and establish consistent communications with the customer on work progress	C4	9	33.3
5	Establishing proactive cost contingency plan	Establish a proactive cost contingency plan with full reference to past projects	D6	7	25.9
6	Setting up a time buffer for regulatory approval	Set up a time buffer for the regulatory approval when planning the project, and avoid errors and omissions in the submitted documents	E3	10	37.0
7	Stipulating strict self-regulatory rules	Stipulate strict self-regulatory rules and conditions to avoid producing pollution during construction	F3	7	25.9
8	Warning of corruption consequence	To warn all site staff of consequences upon discovery of such behavior and be a good role model	G1	1	3.7

project are commercial entities, and different types of commercial entities (e.g., retail sales, food and beverage services, hotel, and office) have different expectations for commercial projects (Zhang, 2014). Disputes arise inevitably when expectations from customers are poorly managed, which would affect the successful deliveries of commercial projects. Moreover, green commercial construction projects are more complicated in construction technologies compared with traditional ones (Korkmaz et al., 2010), which would probably increase the potential for a dispute. According to respondent feedback, this issue has raised significant concerns from the industry, and their normal response measure is to employ a professional consultant with expertise in commercial property to help manage expectations from diverse customers.

Within the financial risk, the mitigation measure with the most recommendations is "establishing proactive cost contingency plan," which addresses the risk factor D6 "additional costs due to the adoption of green material and equipment." Respondents suggested that a proactive and deliberate cost estimate for green material and equipment should be performed and that a comprehensive review of past projects of similar type should be conducted to guarantee estimate accuracy. Such a measure has also been widely recommended by previous studies to address the issue of cost overrun in construction projects (Hwang et al., 2015d; Flyvbjerg, 2007; Zou et al., 2007).

Within the legal risk, the mitigation measure mentioned with the highest frequency is "setting up a time buffer for regulatory approval," which is intended to address the risk factor E3 "delay in regulatory approvals for new green requirements." Respondents suggested that a float period or a time buffer should be preestablished for regulatory approvals when planning the project. They likewise emphasized that errors and omissions in submitted documents should be minimal to ensure that the review by the authority proceeds smoothly.

Within the environmental risk, the mitigation measure mentioned at the highest frequency is "stipulating strict self-regulatory rules," a solution for the risk factor F3 "air, sound, and water pollution." Pollution derived from construction sites is a significant concern of the Singapore construction authority and the local public. Forfeit will be incurred to a construction company if pollution is detected from its work in progress (The Straits Times, 2016). According to respondent feedback, the majority of local construction companies have stipulated strict self-regulatory rules to keep their work process free from producing pollution.

In addition, within the category of political risk, only one mitigation measure, namely, "warning of corruption consequence," was mentioned to address risk factor G1 "corruption and bribery." Such a result is unanticipated but understandable; after all, the political situation in Singapore has been stable for decades, without a sign of any possible change (Kim Cheng Low, 2006). Thus, local construction companies have input quite a few efforts regarding the political risk.

14.6 SUMMARY

The risk management of green construction projects was considered in this chapter. The risks of green retrofit, residential, and commercial construction projects were identified; the criticalities of these risks in green and conventional construction projects were compared; and feasible measures that may mitigate risks were proposed.

For green retrofit projects, the top eight critical risks were "postretrofit tenants' cooperation risk," "regulatory risk," "market risk," "financial risk," "preretrofit tenants' cooperation risk," "different concerns of stakeholders," "material supply and availability," and "quality risk." Results also showed that green retrofit projects face risks at a higher magnitude than conventional retrofit projects, because 19 out of 20 risks received higher RC values in green retrofit projects than in conventional retrofit projects. In addition, this chapter recommended 28 measures that can effectively mitigate the risks in green retrofit projects.

For green residential construction projects, survey results showed that, "complex procedures to obtain approvals," "overlooked initial high cost," "unclear requirements of owners," "employment constraint," and "lack of availability of green materials and equipment," were the top five critical risks. The comparison of risk criticalities between green and conventional projects showed that 35 out of 42 risks gained higher RC values in green projects, implying green residential construction projects face a more challenging situation in risk management. Furthermore, this chapter recommended a series of feasible measures that may mitigate the risks in green residential construction projects, and the top five effective of them were "improve communication and coordination among contracting parties," "understand owner's goal of the Green Mark Standard," "use past successful green residential projects as references," "develop training programs to upgrade workers' skill and knowledge on new technologies and materials," and "allow for contingency funds."

For green commercial construction projects, results showed that the top five critical risks were "inflation," "currency and interest rate volatility worsened by the import of green materials," "durability of green materials," "damages caused by human error," and "shortage of green materials." Results also showed that green commercial construction projects face challenges and risks at a greater magnitude than traditional ones concerning the issues of adoption of green ideas, materials, technologies, and incurred cost. In addition, seven widely used mitigation measures against these risks, namely, "inspecting quality frequently with a detailed checklist," "ensuring sound project design and early briefing," "guaranteeing a rigorous contractor selection," "managing customer expectations soundly," "establishing proactive cost contingency plan," "setting up a time buffer for regulatory approval," and "stipulating strict self-regulatory rules," were proposed.

Chapter 15

Leadership Development in Green Construction Projects

15.1 INTRODUCTION

Leadership has been referred to as one of the main enablers of project success in the construction industry (Ahadzie et al., 2008). Frank (2002) indicated that the project manager has direct influence over 34%–47% of project success. In the construction industry, a project manager can be deemed a leader because he/she has the authority to delegate work tasks to his/her project team and to make important decisions on site. Every project manager has his/her own style of doing work and this depends much on his/her leadership style. In other words, different project managers have different leadership styles, and the leadership style used can either affect the project performance positively or negatively. Leadership has been recognized as one of the essential skills of project managers (Edum-Fotwe and McCaffer, 2000; Odusami, 2002) and this is also the case in green building projects (Hwang and Ng, 2013).

Previous studies on the leadership styles of project managers have mostly been based on conventional construction projects and few have been focused on green projects. As there has been a significant rise in green projects, it is important to identify appropriate leadership styles of project managers in these projects because this will help ensure the smooth progress of the projects and their success.

15.2 LEADERSHIP: THEORIES AND INFLUENCE ON PROJECT PERFORMANCE

15.2.1 Classic Leadership Theories

Robbins (2003) defined leadership as the ability to influence a group toward the achievement of goals. Leadership is different from management because the possession of a formal managerial rank in an organization does not necessarily enable a manager to lead effectively and the ability to influence may arise outside the formal structure of an organization. In general, there have been three categories of leadership theories, namely, trait, behavioral, and contingency theories (Robbins and Coulter, 2007).

Performance and Improvement of Green Construction Projects. https://doi.org/10.1016/B978-0-12-815483-0.00015-6

Trait theories focus on the traits of leaders and posit that personal qualities and characteristics are what differentiate leaders from nonleaders (Robbins, 2003). Some studies have revealed strong evidence for an identifiable set of personality and cognitive traits that are expected to characterize successful leaders. For example, Kirkpatrick and Locke (1991) summarized six key traits: drive, leadership motivation, honesty and integrity, self-confidence, cognitive ability, and knowledge of the business; and Turner (1999) identified seven traits of effective project managers: problem-solving ability, results orientation, energy and initiative, self-confidence, perspective, communication, and negotiating ability. Obviously, trait theories of leadership have some limitations. Traits appear to predict leadership in selective situations (Robbins, 2003) and trait theories involve subjective judgments in determining who is a so-called good or successful leader (Mullins, 2007). Indeed, leadership is the combination of traits, skills, and behaviors that leaders use when they interact with their team members (Achua and Lussier, 2013).

Behavioral theories of leadership posit that specific behavior differentiates leaders from nonleaders (Robbins, 2003). The Iowa Studies established three major leadership styles: the autocratic, the democratic, and the laissez-faire leadership styles (Lewin et al., 1939). Specifically, autocratic leaders provide clear expectations for what should be done, when it should be done, and how it should be done; democratic leaders offered guidance to group members, but encouraged participation of employees in decisions; and laissez-faire leaders offered little or no guidance to group members and left decision-making up to group members.

Additionally, in the Ohio State Studies, which began at Ohio State University in the late 1940s (Stogdill and Coons, 1951), two major dimensions of leadership behavior were found: (1) initiating structure and (2) consideration. Initiating structure refers to the extent to which a leader is likely to define and structure his or her role and those of subordinates in the search for goal achievement, while consideration describes the extent to which a leader is likely to have job relationships characterized by mutual trust and respect for group members' ideas and feelings. The two dimensions are independent of each other and can co-exist at the same time.

Moreover, studies undertaken at the University of Michigan proposed two dimensions of leadership behavior: (1) employee-oriented and (2) production-oriented (Likert, 1961). Employee-oriented leaders, also known as person-oriented leaders, emphasize interpersonal relations and care about the needs of their employees, while production-oriented leaders, also known as task-oriented leaders, tend to be more concerned with the accomplishment of tasks. In this theory, the two dimensions affect each other, meaning that when the leader is highly employee-oriented, he/she would be less production-oriented and vice-versa (Northouse, 2013).

Furthermore, Blake and Mouton (1964) proposed a managerial grid based on the styles of concern for people and concern for production. These styles represent the dimensions of consideration or employee-oriented leadership and initiating structure or employee-oriented leadership, identified from the Ohio State studies and University of Michigan studies. The managerial grid presents five typical leadership styles: authority-compliance (9,1), country-club (1,9), team (9,9), middle-of-the-road (5,5), and impoverished (1,1).

Contingency theories of leadership propose that what makes an effective leader depends on the situation. Certain situations and certain follower types need certain leadership styles (Mills et al., 2009). One of the most popular and widely known contingency models is the situational leadership theory developed by Hersey and Blanchard (1974, 1993), who argued that the right leadership style should be contingent on the ability and willingness of the followers. The situational leadership theory consists of two parts: the leadership style, which is dependent on the directive and supportive behavior, and the development level of followers. Therefore, the situational leadership theory presents four leadership styles: (1) directing (low supportive behavior, high directive behavior, low development of followers); (2) coaching (high supportive behavior, high directive, behavior, moderate development of followers); (3) supporting (high supportive behavior, low directive behavior, moderate development of followers); and (4) delegating (low supportive behavior, low directive behavior, high development of followers).

Another contingency theory is the path-goal theory developed by House (1971, 1996). This theory maintains that it is the job of leaders to assist followers in achieving their goals and to provide the necessary direction and/or support to ensure that their goals are compatible with overall organization objectives. House (1996) identified four leadership styles: (1) the directive leader gives specific guidance of performance to followers; (2) the supportive leader is friendly and shows concern for the needs of followers; (3) the participative leader consults with followers and considers their suggestions before making decisions; and (4) the achievement-oriented leader sets high goals and expects followers to have high-level performance. House (1996) assumed that leaders are flexible and that the same leader can practice different types of behavior at different times in varying situations.

Sixteen characteristics of leadership from previous studies on leadership (Achua and Lussier, 2013; Northouse, 2013; Robbins and Coulter, 2007) are discussed in this chapter. They are elaborated upon in the section of results and discussion.

15.2.2 Influence of Leadership on Project Performance and Success

In the project management area, there have been several studies that investigated the relationship between project managers' leadership style and project performance. Odusami et al. (2003) indicated a significant relationship between project managers' leadership styles and project performance in the Nigerian construction industry. This finding was consistent with that of Mustapha and Naoum (1998), who found a significant association between preferred management styles of the site managers and their level of effectiveness. The role of project managers' leadership has also attracted attention from academia. Lee-Kelley and Leong (2003) reported the influence of project managers' leadership style on their perception of project success. Turner and Müller (2005) performed a comprehensive literature review and recognized project managers' leadership as a success factor in projects. Using both quantitative and qualitative studies, Müller and Turner (2007) confirmed the influence of project managers' leadership style on project success and disclosed that different leadership styles are appropriate for different project types. In addition, conducting a questionnaire survey, Geoghegan and Dulewicz (2008) reported that the British project managers' managerial competency leadership dimensions significantly influenced project success. Moreover, Ozorhon et al. (2008) advocated a transformational leadership style and reported that project managers with a transformational leadership style were positively associated with subordinates' work performance in the Thai construction industry. Furthermore, in the Taiwanese construction industry, Yang et al. (2011) found that increases in project managers' leadership enhanced relationships among team members, while team work was positively related to project performance and success. Because the performance and success of green building projects are crucial for the implementation of green strategies in Singapore, this chapter attempts to identify the critical leadership styles adopted by project managers in the green building projects.

15.3 METHOD AND DATA COLLECTION

A literature review helped obtain a better understanding of leadership styles of project managers and collect information to develop the survey questionnaire used for data collection. A pilot study was conducted with three project managers experienced in green construction to validate the questionnaire. In the final survey questionnaire, the first section captured demographic data of the respondents, their companies, and projects. Then, the characteristics of leadership were presented and the respondents were asked to rate their attitudes toward the adoption of these characteristics in their green construction projects, using a five-point Likert scale (1 = strongly disagree; 3 = neutral; and 5 = strongly agree). According to the responses, it is possible to identify the leadership styles adopted by the project managers.

The population of this survey included all the project managers with experience in green construction. A list of project managers certified with the Green Mark Manager (GMM) and Green Mark Professional (GMP) Schemes was obtained from the BCA, and served as the sampling frame of this survey. These project managers were from either developers or contractors. A total of 90 questionnaires were personally delivered, and the researcher clearly explained to the respondents that they should answer the questions based on their experience in green building projects. Finally, 30 responses were received, representing a response rate of 33%. The response rate was reasonable compared with the norm of 20%–30% with most questionnaire surveys in the construction industry (Akintoye, 2000; Hwang et al., 2015c). Also, statistical analysis may be conducted as the central limit theorem holds true with a sample size greater than 30 (Ling et al., 2009; Ott and Longnecker, 2001).

Although Likert scale data have been recognized as ordinal data, a number of studies advocated that Likert data could be analyzed using parametric statistical methods, such as *t*-test (Hwang et al., 2014; Binder, 1984) and factor analysis (Carifio and Perla, 2008; Shan et al., 2015b; Deng et al., 2014a). Nunnally (1975) believed that the results from parametric methods with ordinal data were recognized as reasonably reliable, while other statistics studies argued that parametric methods for interval variables could be used for ordinal variables. This was because the power and flexibility of parametric methods can outweigh the small biases that they may entail (O'Brien, 1979; Allan, 1976), and conclusions and interpretations of parametric methods might be easier and provide more information (Allen and Seaman, 2007). Therefore, the one-sample *t*-test and exploratory factor analysis (EFA) were used to analyze the data.

Table 15.1 shows the profile of the respondents, their companies, and projects: 37% of the respondents were contractors and the remaining 63% were developers. In addition, 83% of the respondents had over 5 years of experience in the construction industry. Also, 60% had less than 3 years' experience in green construction because the GMS has only been mandatory in the construction industry since 2008, when the Building Control (Environmental Sustainability) Regulations were enacted. Furthermore, the responses were based on 44 green building projects. The majority of these projects were residential and commercial building projects, and worth less than S$100 million.

15.4 ANALYSIS RESULTS AND DISCUSSIONS

15.4.1 Leadership Characteristics in Green Construction Projects

The data on the perception of the respondents on leadership characteristics in green construction projects received a Cronbach's alpha of 0.879, indicating high reliability of the data. A one sample *t*-test was further performed to check whether the respondents significantly agreed on the 16 leadership characteristics. As shown in Table 15.2, the results indicated that 13 characteristics with *P*-values below .05 and mean scores above 3 received significant agreement from the project managers.

Characteristic C13 "strive for work performance and productivity" was ranked top (mean = 4.37), suggesting that work performance and productivity were emphasized by project managers in green construction projects. Project managers take the responsibility for achieving project objectives, and the extent

TABLE 15.1 Profile of Companies, Respondents, and Projects

Characteristics			*N*	%
Companies (*N*=30)	Contractors		11	36.67
	Developer		19	63.33
Respondents (*N*=30)	Experience in green construction	≤3 years	18	60.00
		4–6 years	8	26.67
		>6 years	4	13.33
	Experience in the construction industry	≤5 years	5	16.67
		6–10 years	18	60.00
		11–15 years	4	13.33
		16–20 years	2	6.67
		>20 years	1	3.33
Project (*N*=44)	Project type	Residential	20	45.45
		Commercial	21	47.73
		Mixed development	3	6.82
	Project size (S$)	<20 million	5	11.36
		20–40 million	8	18.18
		40–60 million	16	36.36
		60–80 million	5	11.36
		80–100 million	2	4.55
		>100 million	8	18.18

to which they achieve project objectives is considered as their performance and linked to their bonuses. In some Singaporean firms, project managers have signed accountability pledges, in which they formally announce their responsibility for achieving the expected project performance (Zhao et al., 2014a). In addition, some efforts have been made to develop a standardized performance measurement system for projects in Singapore (Hwang et al., 2015e). Examples are the Buildable Design Appraisal System (BDAS) (BCA, 2005a)

TABLE 15.2 Leadership Styles of Project Managers in Green Construction Projects

Code	Characteristics of Leadership	Mean	Rank	*P*-Value
C01	Make all the decisions without discussing with subordinates	2.50	16	.000*
C02	Tell subordinates clearly what to do and how to do it	4.00	6	.000*
C03	Closely supervise the work of subordinates	4.00	6	.000*
C04	Encourage subordinates to shares their ideas and opinions	3.77	9	.000*
C05	Work closely with subordinates to determine what needs to be done	4.07	4	.000*
C06	Direct subordinates with clear roles and goals	4.10	3	.000*
C07	Have high concern for work tasks	4.13	2	.000*
C08	Be sensitive to the feelings of subordinates	3.43	13	.001*
C09	Look out for subordinates' welfare	3.60	11	.000*
C10	Focus on getting the job done without paying any regard to subordinates' welfare and feelings	2.57	15	.003*
C11	Develop and maintain good relationships with subordinates	3.77	9	.000*
C12	Encourage two-way communication	4.07	4	.000*
C13	Strive for work performance and productivity	4.37	1	.000*
C14	Have high concern for people	3.53	12	.000*
C15	Empower subordinates with the freedom or autonomy in making decisions	3.13	14	.293
C16	Delegate work responsibilities to subordinates	3.93	8	.000*

**The one-sample* t*-test result is significant at the .05 level.*

and the Construction Quality Assessment System (CONQUAS) (BCA, 2005b). Furthermore, it is not surprising that project managers strive for project productivity in Singapore because the BCA has come up with several initiatives, such as the Construction Productivity Roadmap (CPR) and the Constructability Appraisal System (CAS), to enhance construction productivity.

C07 "have high concern for work tasks" received the second position (mean = 4.13), indicating that work tasks were highlighted by project managers in green building projects. This could be due to the fact that projects are highly goal-oriented (Frame, 2003). The high rank of this characteristic was also consistent with the high rank of C13 because both of them were related to project objectives. In reality, no matter in traditional or green building projects, project objectives are highlighted because the achievement of those objectives represents project success in most cases (Ling et al., 2009).

C06 "direct subordinates with clear roles and goals" was ranked third (mean = 4.10), indicating that the project managers tended to direct their subordinates with clear roles and goals so that subordinates know their part and role to play in the project. In addition, with clear roles and goals, subordinates are able to understand what the project manager expects of them. Additionally, in most cases, there are various groups of people working on site in a construction project because different work activities have to be conducted (Turner, 2009). Hence, it is appropriate for the project manager to direct his/her team with clear roles and goals to prevent any confusion.

C05 "work closely with subordinates to determine what needs to be done" occupied the fourth position (mean = 4.07), which was consistent with the democratic leadership style in the University of Iowa Studies, because democratic leaders would like to actively engage their subordinates, seek their participation, and work with them to determine what needs to be done (Achua and Lussier, 2013). In addition, green technologies tend to require complicated techniques and construction processes (Zhang et al., 2011), with which staff may not be familiar. Because of the unfamiliarity of green technologies, a project manager should work more closely with the subordinates to determine the work tasks and clarify any issues (Hwang and Ng, 2013).

C12 "encourage two-way communication" was ranked fourth as well (mean = 4.07), indicating that project managers in green building projects also highlighted two-way communication within project teams. According to the Project Management Institute (PMI, 2008), project communication management has been one of the nine areas of the Project Management Body of Knowledge (PMBOK), which shows the important role of communication in projects. In green building projects, communication skills allow the project managers to disseminate information and work procedures within diverse project participants. Additionally, because of the unfamiliarity with green technologies and complexity of green building projects, project managers should communicate regularly with project participants (Hwang and Ng, 2013) and clarify work tasks, thus reducing the likelihood of erroneous work and rework.

C02 "tell subordinates clearly what to do and how to do it" was ranked sixth (mean = 4.00), suggesting that the survey project managers preferred to use the directive leadership style in the path-goal theory as well as the directing style in the situational leadership theory. The high mean score of this characteristic was consistent with the ordinary practice in construction projects. Project managers should give their subordinates clear instructions on work tasks and construction methods.

Also, C03 "closely supervise the work of subordinates" was ranked sixth (mean = 4.00). Close supervision of a project manager can help ensure that a project is on the right track and that work is being carried out correctly. In Singapore, project managers usually conduct daily site walkabouts to check the work progress. Another reason for the close supervision of project managers was that not everyone in the industry had the relevant knowledge and skills, especially those relating to green technologies. The lack of close supervision from project managers would result in erroneous work and rework.

Moreover, C16 "delegate work responsibilities to subordinates" was ranked eighth (mean = 3.93), suggesting that the surveyed project managers tended to allocate some responsibilities to the subordinates appropriate for the work. C04 "encourage subordinates to shares their ideas and opinions" was ranked ninth (mean = 3.77). This result indicates that the project managers encouraged the participation of their subordinates in decision making, consistent with the participative leadership style in the path-goal theory and democratic leadership style in the University of Iowa Studies. C11 "develop and maintains good relationships with subordinates" also occupied the ninth position (mean = 3.77). This was in accordance with the middle-of-the-road leadership style in the managerial grid model. In the context of Singapore, project staff are usually from different countries with various national cultures (Hwang et al., 2013c). Hence, cultural differences should be highlighted when project managers handle their relationships with subordinates (Low and Shi, 2002).

Furthermore, both C01 and C15 were related to decision-making, with low mean scores. As some respondents revealed, in the Singapore construction industry, project managers had to make decisions all the time. In most cases, project managers did not discuss minor decisions with their subordinates. However, when making major decisions, they held discussions with their subordinates to garner more opinions and viewpoints.

15.4.2 Leadership Grouping in Green Construction Projects

Exploratory factor analysis (EFA) identifies a relatively small and manageable set of underlying factor groupings that can be used to represent the correlations among a large set of interrelated variables. In this chapter, EFA was used to categorize the 13 leadership characteristics that received significant agreement from the surveyed project managers. The ratio of the sample size to the number of variables was 2.31, which is less than the normal threshold of 5. However,

MacCallum et al. (1999) concluded that factor analysis can produce correct solutions even with samples that would traditionally have been determined to be too small for meaningful factor analysis. In addition, some previous studies using the EFA technique had a ratio below 5 [e.g., 2.48 in Koksal and Arditi (2004); 3.39 in Li et al. (2005); and 1.59 in Shen and Liu (2003)].

The appropriateness of the factor analysis for the factor extraction was assessed in various ways. The Kaiser-Meyer-Olkin (KMO) value was 0.745, indicating an acceptable degree of common variance among the characteristics. The value of the test statistic for Bartlett's sphericity was large (Chi-square = 302.908) and the P-value was .000, suggesting that the population correlation matrix was not an identity matrix. Hence, it can be concluded that the data were appropriate for EFA.

As for factor extraction, principal component analysis was used to identify underlying grouped factors. In terms of the number of factors to be extracted, Horn's parallel analysis (Horn, 1965) was adopted because this method has been seen as the most accurate method to determine the number of factors to retain (Pallant, 2010).

Factor rotation can be orthogonal or oblique. In this chapter, the oblique rotation was applied because the orthogonality of factors is often an unrealistic assumption. In almost all areas of social science, any factor is more or less related to other factors, and arbitrarily forcing the factors to be orthogonal may bias the reality. Even if the factors are indeed uncorrelated, the orthogonality should be empirically verified and demonstrated via EFA with an oblique rotation method (Matsunaga, 2010; Zhao et al., 2013b). In this chapter, promax rotation was adopted. The promax method begins with a varimax solution and then raises the factor loadings to a stated power called kappa (κ). Thus, loadings of small magnitude will approach 0 while large loadings, although reduced, are still substantial (Zhao et al., 2015).

The principal component analysis with promax rotation was performed and the result indicated two components with eigenvalues over 1. Then, the two eigenvalues were compared with the criterion values from the parallel analysis using the computer program developed by Patil et al. (2007). The number of components to retain is the number of eigenvalues that are larger than the corresponding random eigenvalues (i.e., criterion values). As Table 15.3 shows, the number of components from the parallel analysis is two, consistent with that from the principal component analysis. The two leadership characteristic groupings explained approximately 68.8% of the variance, higher than the guideline of 60% recommended by Malhotra (2006).

The factor loading value reflects the degree of contribution of individual variables to each underlying grouping. All of the characteristics obtained factor loadings above the threshold of 0.45 (Comrey, 1973). Thus, the 13 characteristics were categorized into two groupings. Each grouping was labelled by considering the characteristics with high loadings and what they had in common.

15.4.2.1 Grouping 1: Directive and Task-Oriented Leadership

The Grouping 1 was interpreted as "directive and task-oriented leadership." The two characteristics with the highest factor loadings were C06 "direct subordinates with clear roles and goals" and C02 "tell subordinates clearly what to do and how to do it." Both of them are consistent with the directive leadership style in the path-goal theory and the directing style in the situational leadership theory. In addition, both of them highlight the role and goals of subordinates. Achieving the goal of each subordinate can collectively contribute to the achievement of the project task.

TABLE 15.3 Leadership Characteristic Groups

		Grouping	
Code	**Characteristics of Leadership**	*1*	*2*
C06	Direct subordinates with clear roles and goals	0.976	
C02	Tell subordinates clearly what to do and how to do it	0.935	
C07	Have high concern for work tasks	0.823	
C13	Strive for work performance and productivity	0.759	
C03	Closely supervise the work of subordinates	0.796	
C16	Delegate work responsibilities to subordinates	0.694	
C05	Work closely with subordinates to determine what needs to be done	0.631	
C14	Have high concern for people		0.887
C11	Develop and maintain good relationships with subordinates		0.885
C08	Be sensitive to the feelings of subordinates		0.880
C09	Look out for subordinates' welfare		0.869
C04	Encourage subordinates to shares their ideas and opinions		0.727
C12	Encourage two-way communication		0.542
Eigenvalue		6.677	2.269
Criterion value from parallel analysis[a]		2.337	1.929
Variance (%)		51.4%	17.4%
Grouping mean		4.09	3.70

Grouping 1 = Directive and task-oriented leadership; Grouping 2 = Relationship-oriented leadership.
[a]*The criterion values were obtained via the computer program developed by Patil et al. (2007). The "number of variables" = 13 (i.e., the number of leadership characteristics). Sample size = 30.*

C07 "have high concern for work tasks," C13 "strive for work performance and productivity," and C03 "closely supervise the work of subordinates" are indeed task-oriented leadership. In other words, the concern for work task, performance, and productivity indicates that leaders need to pay great attention to the final fulfillment of project objectives. To ensure that a project is performed smoothly, project managers tend to closely supervise the work of subordinates. The close supervision of project managers can also represent their high concern for the work tasks as well as project performance and success.

To ensure project success, project managers also need to encourage the participation of their subordinates. Thus, C16 "delegate work responsibilities to subordinates" and C05 "work closely with subordinates to determine what needs to be done" were also included in Grouping 1. It should be noted that the participation of the subordinates does not mean that they are empowered to make decisions, as suggested by the relatively low mean score of C15 "empower subordinates with the freedom or autonomy in making decisions."

15.4.2.2 Grouping 2: Relationship-Oriented Leadership

The Grouping 2 was labelled as "relationship-oriented leadership." Four leadership characteristics (C08, C09, C11, and C14) received high factor loadings over 0.800. C14 "have high concern for people" received the highest factor loading. High concern for the subordinates contributes to the development of relationships within the project team. C08 "be sensitive to the feelings of subordinates" and C09 "look out for subordinates' welfare" can be seen as specific examples of high concern for subordinates. Thus, a good relationship between a project manager and subordinates can be gradually developed and maintained. These four leadership characteristics are in accordance with the supportive leadership style in the path-goal theory.

C04 "encourage subordinates to shares their ideas and opinions" was also in Grouping 2. This characteristic aligns with the participative leadership style in the path-goal theory and democratic leadership style in the University of Iowa studies. Encouraged by project managers, subordinates would like to share their ideas and opinions about projects in which they are engaged.. This also indicates that project managers respect the ideas and opinions from subordinates, which contributes to the relationship between project managers and subordinates. Additionally, C12 "encourage two-way communication" was included in Grouping 2 because communication could help resolve disputes or disagreements, contributing to a good relationship.

15.4.2.3 Comparison Between the Two Groupings

As shown in Table 15.3, the mean scores of Groupings 1 and 2 are 4.09 and 3.70, respectively. The mean score of a grouping was calculated as the average value of all the mean scores of the leadership characteristics under this grouping.

The result suggested that the leadership style of project managers in green building projects was more toward "directive and task-oriented leadership." It had been only 5 years since the GMS became mandatory in Singapore's construction industry. Hence, it was not uncommon that many staff were still unfamiliar with green technologies and construction processes, which were usually more complicated than those of traditional projects (Zhang et al., 2011; Hwang and Ng, 2013). The project managers, who were certified with the GMM and GMP Schemes and knowledgeable about the green technologies and construction processes, should provide directions for their subordinates and try to avoid any erroneous work. In addition, the BCA has set up a target that at least 80% of buildings will have been certified by the GMS by 2030 (BCA, 2009). Thus, the green building projects under construction should be completed successfully, which contributes to the achievement of this target. Project managers, who are accountable for their projects, should highlight the achievement of project objectives, thus being more task-oriented.

However, the project managers did not overlook their subordinates' roles. The project managers still cared about the feelings, welfare, and opinions of their subordinates. When dealing with their relationships with the subordinates, the project managers adopted participative and democratic leadership styles. Indeed, a harmonious relationship within a project can greatly contribute to the project success (Chua et al., 1999).

Therefore, the project managers in green building projects should be not only directive and task-oriented, but also relationship-oriented. These two leadership styles should not be separated because the relationships within a project team inevitably influence project performance and productivity, while the interaction between a project manager and subordinates facilitates the establishment of relationships.

15.5 SUMMARY

Leadership has been recognized as one of the essential skills of green construction project managers (Hwang and Ng, 2013) and it is crucial to the success of these projects. The objective of this chapter was to identify the leadership characteristics and styles of project managers in green building projects in Singapore. The analysis results indicated that 13 leadership characteristics obtained significant agreement from the project managers in the Singaporean green construction projects. Among these characteristics, "strive for work performance and productivity," "have high concern for work tasks," and "direct subordinates with clear roles and goals," occupied the top three positions. Additionally, the 13 characteristics were categorized into two groupings via EFA: (1) directive and task-oriented leadership and (2) relationship-oriented leadership. The leadership style of the project managers was found to lean more toward directive and task-oriented leadership. However, project managers should not neglect their relationships with subordinates because these relationships also influence project performance and productivity.

Chapter 16

Conclusions and Recommendations

16.1 CONCLUSIONS

Over recent years, green buildings have already achieved rapid development worldwide owing to a global recognition of their environment-friendly benefits. Accordingly, green construction has also grown dramatically and has become an important and promising subsector of the current construction industry. Unfortunately practitioners have indicated that green building projects do not always run as smoothly as expected. It appears that green construction is facing various management challenges, such as massive cost overruns, significant delivery delay, low construction productivity, increasing workplace injuries, and extensive reworks. As a result, this book has sought to look into these management issues and has tried to provide some feasible and effective solutions that can tackle these issues and challenges.

Chapter 1 is a general introduction to the book, in which the research motivation, research scope, research objectives, research significance, and the structure of the book, were sequentially introduced. Chapter 2 provided the background of this book, in which the conception of green construction projects, the characteristics of green construction projects, as well as a comprehensive overview of the green construction industry of Singapore were carefully explained. Chapter 3 concerns methodology, in which the research design, data collection and data analysis approaches were elaborated.

Chapter 4 presented a comprehensive review of the 15 prevailing green building rating systems (GBRSs) used in the world. The review showed that BREEAM from the United Kingdom, LEED from the United States, and BEAM from Hong Kong were the early GBRSs launched in the 1990s, while the rest of the identified GBRSs were primarily released in the first decade of the 21st century. The review also showed that the prevailing GBRSs offered a wide variety of particular versions in terms of project stages, project types, and building types. Furthermore, the review identified seven essential evaluation criteria that are commonly used by the prevailing GBRSs, and found that the most important evaluation criterion was "energy," following by "site," "indoor environment," "land and outdoor environment," "material," "water," and "innovation." This chapter also revealed that the majority of the identified GBRSs were using a

Performance and Improvement of Green Construction Projects. https://doi.org/10.1016/B978-0-12-815483-0.00016-8

four-level hierarchy rating system or above to rate and certify green buildings. Additionally, this chapter identified four major themes within the current GBRS research, which were "comprehensive review of GBRSs used worldwide," "reviews and discussions of regional GBRSs," "comparison between different GBRSs on particular aspects," and "the development of particular GBRSs tailored to certain purposes." Furthermore, this chapter also pointed out three directions that the future GBRS research may go toward, namely, "the investigation of the barriers and drivers for using GBRS," "the examination of project stakeholders' perceptions of GBRSs," and "the development of a corporate sustainability rating system."

Chapter 5 examined the obstacles hindering the promotion of green construction projects and found the top five obstacles to be "high premium costs," "the lack of communication and interest between project members," "the lack of expressed interest from clients or market demand," "the lack of credible research on the benefits of green buildings," and "green building practices are costly to implement." Furthermore, Chapter 5 proposed three suites of solutions in order to overcome these obstacles. The proposed solutions were "providing governmental incentives for green construction projects," "improving the communication within the project team," and "conducting research on benefits of green buildings."

Chapter 6 examined the key knowledge areas and skills required for performing green construction projects. Results showed that the most important knowledge areas were "schedule management and planning," "stakeholder management," "communication management," "cost management," and "human resources management." The most important skills were analytical, decision making, teamwork, delegation, and problem-solving skills.

Chapter 7 undertook a systematic review of the practices, polices, initiatives, and research efforts in the area of green construction project financing. It reviewed the green construction project financing practices implemented in representative developed economies, such as the United Kingdom, the United States, Singapore, and Australia. It also reviewed the efforts and initiatives of green construction launched by selected international organizations, such as the UN, OECD, and IFC. Furthermore, this chapter reviewed the research efforts conducted by the existing green construction project financing studies, and identified four major research themes within the current green construction project financing research, namely, "the review of financial stakeholders and market of green construction," "the benefits and barriers to green construction project financing," "the identification of financial vehicles for green construction," and "the development of innovative models and mechanisms for green construction project financing." This chapter also identified four future directions for the green construction project financing research, which were "the identification of financial issues in green construction projects," "the examination of critical drivers for implementing green construction project financing," "the investigation of financial vehicles for green construction projects," and "the development of best practices for implementing green construction project financing."

In spite of the benefits of green buildings and considerable efforts made to promote a sustainable built environment, the deliveries of green buildings are still hindered by the higher cost associated with "going green." Chapter 8 of the book revealed that the existing green cost premiums were 5%–10%, with green residential buildings having the highest cost premiums, followed by green commercial and green office buildings. It was also demonstrated in Chapter 8 that project type and project size were statistically significant variables that affected cost premiums. Furthermore, Chapter 8 identified that "high cost of green technologies and materials," "higher research and development costs for green building products, systems, technologies etc.," and "lack of required green expertise and information" were the top three reasons for the cost premiums of green construction projects. As for the current cost performance, Chapter 8 concluded that green construction projects were generally over budget by 4.5%–7%, which was worse than conventional construction projects. As for the solutions that can reduce cost premiums and improve the cost performance of green construction projects, Chapter 8 revealed that "tax relief" would be the most effective one.

Chapter 9 of the book examined the schedule performance of green construction projects and found that, on average, green construction projects were delayed by 4.8% when compared with their as-planned schedules. Furthermore, the top five factors causing delay in green construction projects were revealed as "speed of decision making by client," "speed of decision-making involving all project teams," "communication/coordination between key parties," "level of experience of consultants," and "difficulties in financing project by contractors."

Construction safety is another major issue facing green construction projects, and it has been found that green construction projects recorded a worse safety performance than conventional construction projects. Chapter 10 identified the top 10 critical safety issues of green construction projects, which were "fall from height," "electrocution," "being struck by falling objects," "contact with/by machinery," "fall on same level from slipping, tripping, and stumbling," "contact with hand-tools," "contact with heavy vehicles," "crush injuries," "fall on same level due to collision with, or pushing by, another person," and "moving/handling heavy loads." Also, 10 feasible solutions to address these safety issues were recommended.

Chapter 11 of the book presented a detailed investigation of the rework issues in green construction projects. Results showed that, although the incidence of rework in green construction projects was lower than that of conventional construction projects, the adverse impacts of rework on green construction projects were higher than those on conventional projects. Chapter 11 also unveiled the top four critical rework factors in green construction projects, which were "owner change," "design change," "design error/

omission," and "contractor's error/omission." In addition, Chapter 11 recommended five practical solutions, namely, "use communication tools for design coordination," "ensure the correctness of design," "ensure the compatibility of design and specifications," "establish an effective channel for communication," and "ensure a sound change management," to reduce the reworks in green construction projects.

Chapter 12 of the book looked into the construction productivity of green construction projects. It revealed that only 22.0% of conventional construction projects had encountered productivity issues, while green construction projects almost doubled that figure to 42.8%. Furthermore, Chapter 10 identified the top five factors affecting the productivity of green construction projects, which were "workers' experience," "design changes," "technology," "workers' skill level," and "planning and sequencing of work." In addition, Chapter 12 presented the top three useful solutions that can improve the productivity of green construction projects, which were "standardized design," "investment in new tools and equipment," and "consultants' or superintendents' on time checking, receiving and approving of design documents."

Chapter 13 of the book developed an assessment model to gauge the risks in green construction projects, with the aid of the fuzzy set theory approach. Using the developed model, the likelihoods of occurrence, magnitudes of impact, and risk criticalities of the risk factors, groups and the overall risk in green construction projects were calculated. The results indicated that "inaccurate cost estimation," "delay in issuance of documents," and "unclear detailed design or specifications," were the top three critical risk factors in green construction projects. At the risk group level, it reported that "cost overrun risk," "client-related risk," and "procedure complexity" were the top three risk groups. Furthermore it showed that the overall risk criticality calculated by the assessment model was high, indicating risk management is still necessary for green construction in Singapore.

In addition to developing the risk assessment model, the critical risks in green retrofit, residential, and commercial construction projects were investigated in Chapter 14. Results showed that the top eight critical risks to green retrofit projects were "postretrofit tenants' cooperation risk," "regulatory risk," "market risk," "financial risk," "preretrofit tenants' cooperation risk," "different concerns of stakeholders," "material supply and availability," and "quality risk." The top five critical risks to green residential construction projects were "complex procedures to obtain approvals," "overlooked initial high cost," "unclear requirements of owners," "employment constraint," and "lack of availability of green materials and equipment." Finally, the top five critical risks to green commercial construction projects were "inflation," "currency and interest rate volatility worsened by the import of green materials," "durability of green materials," "damages caused by human error," and "shortage of green materials." A series of practical solutions to tackle these identified risks were also suggested.

Chapter 15 of the book identified the leadership characteristics and the styles of project managers in existing green construction projects. Results showed that, "striving for work performance and productivity," "having high concern for work tasks," and "directing subordinates with clear roles and goals," were the top three leadership characteristics of green project managers. Moreover, the leadership style of the green project managers was found to lean more toward directive and task-oriented leadership. It was also noted that green project managers should not neglect their relationships with subordinates, because those relationships also influenced project performance and productivity.

16.2 CONTRIBUTIONS

The first major contribution of this book is in providing academic researchers with a series of conceptual frameworks to manage the various issues and challenges encountered in the implementation of green construction projects. Although considerable research effort is being devoted to the area of green construction nowadays, little of it examines the management challenges associated with the implementation of green construction projects. This has resulted in a lack of mature management theories for academic researchers to refer to when they are investigating the management problems of green construction projects. As presented from Chapters 4 to 13, this book has developed a series of theoretical frameworks that address various management challenges in green construction projects, such as cost, schedule, safety, quality, risk, productivity, and leadership, and thus has contributed to the current body of knowledge of green construction.

In addition to its contribution to academia, this book can contribute to the practice as well. It has reviewed the existing green construction financing initiatives and green building rating systems, examined the management issues, problems, and risks of green construction projects facing industry practitioners, and investigated how different green construction projects are from conventional construction projects in the sense of management performance and approach. This book is informative and can also serve as a helpful manual or a handbook for industry professionals working on green construction projects. Relying on the knowledge gained from this book, industry practitioners can enhance their understanding of, and capabilities in, managing green construction projects.

This book is also helpful to policymakers. It has provided policymakers with a full picture of the issues, obstacles, challenges, barriers, and risks of green construction projects, thus enhancing their understanding of such projects. Furthermore, as the main body of this book was derived from empirical surveys of industry practitioners, it has also provided policy makers with a good opportunity to learn about the industry practitioners' perceptions of green construction projects, thereby allowing them to develop more effective policies.

16.3 LIMITATIONS

Despite the detailed investigation of the management problems in green construction projects, some limitations are still present in this book. First, the sample sizes of the surveys mentioned in this book are relatively small (30 or so). Thus, caution should be given when the analysis results are interpreted and generalized. Second, this book collected opinion-based data from industry practitioners, which is subjective to a certain extent, and bias may occur owing to the experience and preference of respondents. Lastly, the findings from this book are based on the context of Singapore exclusively, and therefore they may vary in other countries.

16.4 RECOMMENDATIONS

Green construction is a promising area worthy of continuous research input and there have been several clear directions for future research. First, the interrelationships among the risks in green construction projects can be investigated as this can provide an in-depth understanding of the risks of such projects for both academia and practitioners. But more than that, a risk allocation model can also be proposed to allocate the risks among different stakeholders. Moreover, a construction time prediction model can be developed to predict the duration of green construction projects, and it is also recommended to establish the entire process of project planning and feasibility studies for green projects as the processes directly affect schedule performance. Another avenue for future studies would be development of a benchmarking system to assess the performance of green construction projects, allowing the practitioners to compare their performance with a reference point and helping them learn from leading organizations through the adoption of best practices. Additionally, it would be interesting to conduct in-depth case studies to disclose how project managers lead green building projects and compare their leadership styles with those of conventional projects. Also, it is recommended to identify the factors affecting the leadership styles in green construction projects and to investigate the theoretical rationale behind these factors using organizational behavior theories.

References

Abdel-Razzaq, J., 2016. Half of Saudi Green Building Projects Facing Delays. Available from: https://www.thebig5hub.com/news/2016/february/half-of-saudi-green-building-projects-facing-delays/ (Accessed 23 February 2017).

Abu Dhabi Urban Planning Council, EOAD, 2010. The Pearl Rating System for Estidama: Building Rating System. Available from: https://www.upc.gov.ae/en/estidama/pearl-rating-system/pearl-building-rating-system (Accessed 10 March 2017).

Abu-Bakar, S.H., Muhammad-Sukki, F., Ramirez-Iniguez, R., Mallick, T.K., Mclennan, C., Munir, A.B., Mohd Yasin, S.H., Abdul Rahim, R., 2013. Is renewable heat incentive the future? Renew. Sust. Energ. Rev. 26, 365–378.

Abudayyeh, O., Fredericks, T.K., Butt, S.E., Shaar, A., 2006. An investigation of management's commitment to construction safety. Int. J. Proj. Manag. 24, 167–174.

Achua, C.F., Lussier, R.N., 2013. Effective Leadership. Cengage South-Western, Florence, United States.

Addis, B., Talbot, R., 2001. Sustainable Construction Procurement: A Guide to Delivering Environmentally Reponsible Projects. Construction Industry Research & Information Association (CIRIA), London.

Agarwal, S., Satyanarain, R., Sing, T.F., Vollmer, D., 2016. Effects of construction activities on residential electricity consumption: evidence from Singapore's public housing estates. Energy Econ. 55, 101–111.

Aghazadeh, F., Mital, A., 1987. Injuries due to handtools: results of a questionnaire. Appl. Ergon. 18, 273–278.

Ahadzie, D.K., 2007. A Model for Predicting the Performance of Project Managers in Mass House Building Projects in Ghana. University of Wolverhampton, Wolverhampton, United Kingdom.

Ahadzie, D.K., Proverbs, D.G., Olomolaiye, P., 2008. Towards developing competency-based measures for construction project managers: should contextual behaviours be distinguished from task behaviours? Int. J. Proj. Manag. 26, 631–645.

Akintoye, A., 2000. Analysis of factors influencing project cost estimating practice. Constr. Manag. Econ. 18, 77–89.

Akintoye, A.S., Macleod, M.J., 1997. Risk analysis and management in construction. Int. J. Proj. Manag. 15, 31–38.

Aladağ, H., Işik, Z., 2017. Role of financial risks in BOT megatransportation projects in developing countries. J. Manag. Eng. 33.

Alaghbari, W., Kadir, M.R.A., Salim, A., Ernawati, 2007. The significant factors causing delay of building construction projects in Malaysia. Eng. Constr. Archit. Manag. 14 (2), 192–206.

Al-Bahar, J.F., Crandall, K.C., 1990. Systematic risk management approach for construction projects. J. Constr. Eng. Manag. 116, 533–546.

Alhorr, Y.M., 2017. GSAS Technical Guide 2017. Available from: http://www.gord.qa/admin/content/link2322017121232.pdf (Accessed 10 January 2017).

Al-Humaidi, H., Tan, F.H., 2009. Mobile crane safe operation approach to prevent electrocution using fuzzy-set logic models. Adv. Eng. Softw. 40, 686–696.

Ali, H.H., Al Nsairat, S.F., 2009. Developing a green building assessment tool for developing countries—case of Jordan. Build. Environ. 44, 1053–1064.

Allan, G.B., 1976. Ordinal-scaled variables and multivariate analysis: comment on hawkes. Am. J. Sociol. 81, 1498–1500.

Allen, I.E., Seaman, C.A., 2007. Likert scales and data analyses. Qual. Prog. 40, 64.

Al-Momani, A.H., 2000. Construction delay: a quantitative analysis. Int. J. Proj. Manag. 18, 51–59.

Al-Sabah, R., Menassa, C.C., Hanna, A., 2014. Evaluating impact of construction risks in the Arabian gulf region from perspective of multinational architecture, engineering and construction firms. Constr. Manag. Econ. 32, 382–402.

Ameyaw, E.E., Hu, Y., Shan, M., Chan, A.P.C., Le, Y., 2016. Application of Delphi method in construction engineering and management research: a quantitative perspective. J. Civ. Eng. Manag. 22 (8), 991–1000.

ANDI, 2006. The importance and allocation of risks in Indonesian construction projects. Constr. Manag. Econ. 24, 69–80.

Aneziris, O., Papazoglou, I.A., Mud, M., Damen, M., Kuiper, J., Baksteen, H., Ale, B., Bellamy, L., Hale, A.R., Bloemhoff, A., 2008. Towards risk assessment for crane activities. Saf. Sci. 46, 872–884.

Aneziris, O., Papazoglou, I., Mud, M., Damen, M., Bellamy, L., Manuel, H., Oh, J., 2014. Occupational risk quantification owing to falling objects. Saf. Sci. 69, 57–70.

Arain, F.M., Low, S.P., 2006. Developers' views of potential causes of variation orders for institutional buildings in Singapore. Archit. Sci. Rev. 49, 59–74.

Architecture Week, 2001. Barriers to Building Green. Architecture Week, USA. Available from:http://www.architectureweek.com/2001/0822/environment_1-2.html (Accessed 10 January 2015).

Aritua, B., Smith, N.J., Bower, D., 2011. What risks are common to or amplified in programmes: evidence from UK public sector infrastructure schemes. Int. J. Proj. Manag. 29, 303–312.

Asdrubali, F., Buratti, C., Cotana, F., Baldinelli, G., Goretti, M., Moretti, E., Baldassarri, C., Belloni, E., Bianchi, F., Rotili, A., Vergoni, M., Palladino, D., Bevilacqua, D., 2013. Evaluation of green buildings' overall performance through in situ monitoring and simulations. Energies 6, 6525–6547.

Ashuri, B., Durmus-Pedini, A., 2010. An overview of the benefits and risk factors of going green in existing buildings. Int. J. Facil. Manag. 1 (1), 1–15.

Assaf, S.A., Al-Hejji, S., 2006. Causes of delay in large construction projects. Int. J. Proj. Manag. 24, 349–357.

Assaf, S., Nour, M., 2015. Potential of energy and water efficiency improvement in Abu Dhabi's building sector—analysis of Estidama Pearl Rating System. Renew. Energy 82, 100–107.

Assaf, S.A., Al-Khalil, M., Al-Hazmi, M., 1995. Causes of delay in large building construction projects. J. Manag. Eng. 11, 45–50.

Awadh, O., 2017. Sustainability and green building rating systems: Leed, Breeam, GSAS and Estidama critical analysis. J. Build. Eng. 11, 25–29.

Aye, L., Hes, D., 2012. Green building rating system scores for building reuse. J. Green Build. 7, 105–112.

Azevedo, R., Martins, C., Teixeira, J.C., Barroso, M., 2014. Obstacle clearance while performing manual material handling tasks in construction sites. Saf. Sci. 62, 205–213.

Badu, E., Owusu-Manu, D.-G., Edwards David, J., Holt Gary, D., 2013. Analysis of strategic issues underpinning the innovative financing of infrastructure within developing countries. J. Constr. Eng. Manag. 139, 726–737.

Baker, S., Ponniah, D., Smith, S., 1999. Risk response techniques employed currently for major projects. Constr. Manag. Econ. 17, 205–213.

Barrett, P., Zhang, Y., Moffat, J., Kobbacy, K., 2013. A holistic, multi-level analysis identifying the impact of classroom design on pupils' learning. Build. Environ. 59, 678–689.

BBC, 2015. Green Deal Funding to End, Government Announces. Available from: http://www.bbc.com/news/uk-england-gloucestershire-33638903 (Accessed 14 June 2017).

BCA, 2005a. The Code of Practice on Buildable Design. Building and Construction Authority, Singapore. Available from: http://www.bca.gov.sg/buildabledesign/others/copbdsep05.pdf (Accessed 9 February 2013).

BCA, 2005b. Conquas 21. Singapore. Available from: http://www.bca.gov.sg/professionals/iquas/others/con21.pdf (Accessed 6 January 2013).

BCA, 2009. 2nd Green Building Masterplan. Building and Construction Authority, Singapore. Available from: http://www.bca.gov.sg/greenmark/others/gbmp2.pdf (Accessed 18 October 2014).

BCA, 2010a. Building Planning and Massing. Building and Construction Authority, Singapore. Available from: https://www.bca.gov.sg/greenmark/others/bldgplanningmassing.pdf (Accessed 21 January 2016).

BCA, 2010b. Existing Building Retrofit. Building and Construction Authority, Singapore.

BCA, 2010c. Green and Gracious: Builder Guide. Building and Construction Authority, Singapore.

BCA, 2010d. BCA Green Mark for New Residential Buildings. TBAC Authority, Singapore.

BCA, 2011. Construction Productivity Roadmap to Chart Transformation of the Construction Sector. BCA, Singapore. Available from: http://www.bca.gov.sg/newsroom/pr03032011_cp.html (Accessed 21 January 2016).

BCA, 2012. Landmark Legislation to Green Existing Buildings Under Building Control Act. Building and Construction Authority, Singapore. Available from: http://www.bca.gov.sg/newsroom/pr10092012_bca.html (Accessed 7 January 2014).

BCA, 2013a. $100 Million Green Mark Incentive Scheme for Existing Buildings (GMIS-EB). Building and Construction Authority, Singapore. Available from: http://www.bca.gov.sg/greenmark/gmiseb.html (Accessed 12 January 2014).

BCA, 2013b. BCA Green Mark for New Non-Residential Buildings. Building and Construction Authority, Singapore. Available from: http://www.bca.gov.sg/greenmark/others/gm_nonresi_v4.1_rev.pdf (Accessed 18 October 2014).

BCA, 2013c. Code for Environmental Sustainability for Buildings, third ed. Building and Construction Authority, Singapore.

BCA, 2013d. Productivity Improvement Project (PIP). Building and Construction Authority, Singapore. Available from: http://www.bca.gov.sg/pip/pip.html (Accessed 7 January 2014).

BCA, 2014a. 3rd Green Building Masterplan. Building and Construction Authority, Singapore. Available from: http://www.bca.gov.sg/greenmark/others/3rd_green_building_masterplan.pdf (Accessed 18 October 2014).

BCA, 2014b. BCA Green Mark Assessment Criteria and Online Application. Building and Construction Authority, Singapore. Available from: http://www.bca.gov.sg/greenmark/green_mark_criteria.html (Accessed 2 June 2014).

BCA, 2014c. Construction Demand for 2014 to Remain Strong. Building and Construction Authority, Singapore. Available from: http://www.bca.gov.sg/newsroom/pr09012014_bca.html (Accessed 2 January 2014).

BCA, 2014d. BCA Building Energy Benchmarking Report 2014. Building and Construction Authority, Singapore. Available from: https://www.bca.gov.sg/GreenMark/others/BCA_BEBR_Abridged_FA.pdf.

BCA, 2015a. Contractors Registry. Building and Construction Authority, Singapore. Available from: http://www.bca.gov.sg/contractorsregistry/contractors_tendering_limits.html (Accessed 10 March 2015).

BCA, 2015b. Public Sector Projects to Sustain Construction Demand in 2015. Building and Construction Authority, Singapore. Available from: http://www.bca.gov.sg/newsroom/pr08012015_bca.html (Accessed 4 April 2016).

BCA, 2015c. The Second Construction Productivity Roadmap. Building and Construction Authority (BCA), Singapore. Available from: http://www.bca.gov.sg/emailsender/buildsmart-062015/microsite/others/bca-newsletter.pdf (Accessed 21 January 2016).

BCA, 2015d. BCA Academy-Courses-Certification Courses for Professionals/Specialists. Available from: https://www.bcaa.edu.sg/what-we-offer/courses/certification-courses (Accessed 26 August 2016).

BCA, 2015e. Environment and Property Prospects. BCA, Singapore. Available from: http://www.bca.gov.sg/newsroom/ceosp08012015_bca.html (Accessed 4 April 2016).

BCA, 2015f. Home-Building Control & Management-Legislation on Environmental Sustainability for Buildings. Available from: https://www.bca.gov.sg/envsuslegislation/environmental_sustainability_legislation.html (Accessed 24 March 2016).

BCA, 2015g. Home-Technology-BCA Green Mark Scheme-Enhanced SGD 20 Million Green Mark Incentive Scheme for New Buildings (GMIS-NB). Available from: http://www.bca.gov.sg/greenmark/gmis.html (Accessed 4 May 2016).

BCA, 2015h. Home-Technology-BCA Green Mark Scheme-SGD 100 Million Green Mark Incentive Scheme for Existing Buildings (BMIS-EB). Building and Construction Authority. Available from: https://www.bca.gov.sg/greenmark/gmiseb.html (Accessed 4 May 2016).

BCA, 2016a. BCA Green Mark Assessment Criteria and Online Application. BCA, Singapore. Available from: http://www.bca.gov.sg/greenmark/green_mark_criteria.html (Accessed 4 January 2016).

BCA, 2016b. Green Mark Manager/Green Mark Professional/Green Mark Facilities Manager/Green Mark Facilities Professional. BCA, Singapore. Available from: https://www.bca.gov.sg/greenmark/gm_manager.html (Accessed 4 April 2016).

BCA, 2016c. About Green Mark. Available from: http://www.bca.gov.sg/green_mark/about/about.aspx (Accessed 4 January 2016).

BCA, 2016d. BCA Directory of Registered Contractors & Licensed Builders. Available from: https://www.bca.gov.sg/bcadirectory/classification/details/128 (Accessed 30 July 2016).

BCA, 2016e. Certified Green Mark Manager/Green Mark Professional/Green Mark Facilities Manager/Green Mark Facilities Professional. BCA, Singapore. Available from: https://www.bca.gov.sg/greenmark/gm_manager.html (Accessed 21 November 2016).

BCA, 2016f. Construction Statistics-Construction Output. Available from: https://www.bca.gov.sg/infonet/constat.asp (Accessed 28 August 2016).

BCA, 2016g. Singapore: Leading the Way for Green Buildings in the Tropics. Building and Construction Authority, Singapore.

Beam Society Limited, 2012. Beam Plus New Buildings. Available from: https://www.beamsociety.org.hk/files/download/download-20130724174420.pdf (Accessed 29 September 2017).

Beheiry, S.M., Chong, W.K., Haas, C.T., 2006. Examining the business impact of owner commitment to sustainability. J. Constr. Eng. Manag. 132, 384–392.

Berardi, U., 2013. Stakeholders' influence on the adoption of energy-saving technologies in Italian homes. Energy Policy 60, 520–530.

Bernard, H.R., 2012. Social Research Methods: Qualitative and Quantitative Approaches. Sage, London, United Kingdom.

Bill, A., Roger, T., 2001. Sustainable Construction Procurement: A Guide to Delivering Environmentally Responsible Projects. Construction Industry Research & Information Association, London, UK.

Binder, A., 1984. Restrictions on statistics imposed by method of measurement: some reality, much mythology. J. Crim. Just. 12, 467–481.

Blake, R.R., Mouton, J.S., 1964. The Managerial Grid: Key Orientations for Achieving Production Through People. Gulf, Houston, TX.

Bordass, B., 2000. Cost and value: fact and fiction. Build. Res. Inf. 28, 338–352.

Bortolini, L., Cividino, S.R.S., Gubiani, R., Cecchini, M., Delfanti, L.M.P., Colantoni, A., 2016. Urban green spaces activities: a preparatory groundwork for a safety management system. J. Saf. Res. 56, 75–82.

Bose, A., Ramji, A., Singh, J., Dholakia, D., 2012. A case study for sustainable development action using financial gradients. Energy Policy 47, 79–86.

Bourgeois, M., Breaux, K., Chiasson, M., Mauldin, S., 2010. Tax incentives of going green. CPA J. 80 (11), 19.

Bozovic-Stamenovic, R., Kishnani, N., Tan, B.K., Prasad, D., Faizal, F., 2016. Assessment of awareness of green mark (GM) rating tool by occupants of GM buildings and general public. Energy Build. 115, 55–62.

BRE Global Ltd., 2017. Breeam International New Construction 2016 Technical Manual SD 233 2.0. Available from: http://www.breeam.com/filelibrary/technical%20manuals/breeam_international_nc_2016_technical_manual_2.0.pdf (Accessed 10 April 2017).

Builder, 2009. Construction Tours Promote Green Building. Available from: http://www.builderonline.com/building/building-science/construction-tours-promote-green-building_o (Accessed 1 August 2017).

Building and Construction Authority, 2017. Green Mark for Residential Buildings: 2016 Criteria. Available from: https://www.bca.gov.sg/greenmark/others/gm_rb_2016_criteria_final.pdf (Accessed 10 April 2017).

Building Research Establishment (BRE) & Cyril Sweett, 2005. Putting a Price on Sustainability. BRE Electronic Publications, Bracknell, UK.

Buildings Department of Hong Kong, 2006. Comprehensive Environmental Performance Assessment Scheme for Buildings. Hong Kong, China.

Burleson, R.C., Haas, C.T., Tucker, R.L., Stanley, A., 1998. Multiskilled labor utilization strategies in construction. J. Constr. Eng. Manag. 124, 480–489.

Busch, T., Bauer, R., Orlitzky, M., 2016. Sustainable development and financial markets: old paths and new avenues. Bus. Soc. 55, 303–329.

California Sustainability Alliance, 2009. Greening California's Leased Office Space: Challenges and Opportunities. Available from: http://sustainca.org/sites/default/files/greenleases_report_050509.pdf (Accessed 9 September 2017).

Cameron, I., Hare, B., Davies, R., 2008. Fatal and major construction accidents: a comparison between Scotland and the rest of Great Britain. Saf. Sci. 46, 692–708.

Carifio, J., Perla, R., 2008. Resolving the 50-year debate around using and misusing likert scales. Med. Educ. 42, 1150–1152.

Carpenter, S., 2009. Do green buildings really save energy? Building 59 (2), 22.

CBRE, 2011. Retrofitting Existing Buildings: The Low Cost, High Volume Solution to Climate Change. CB Richard Ellis, Inc., Los Angeles, United States.

Ceran, T., Dorman, A.A., 1995. The complete project manager. J. Archit. Eng. 1, 67–72.

Chalabi, A.F., Camp, D., 1984. Causes of delays and overruns of construction projects in developing countries. CIB Proceedings, pp. 723–734.

Chan, P., 2002. In: Factors affecting labour productivity in the construction industry. Proceedings of the Eighteenth Annual ARCOM Conference. University of Northumbira, Association of Researchers in Construction Management, pp. 771–780.

Chan, D.W.M., Kumaraswamy, M.M., 1996. An evaluation of construction time performance in the building industry. Build. Environ. 31, 569–578.

Chan, E.H., Qian, Q.K., Lam, P.T., 2009. The market for green building in developed Asian cities—the perspectives of building designers. Energy Policy 37, 3061–3070.

Chan, A.P.C., Yeung, J.F.Y., Yu, C.C.P., Wang, S.Q., Ke, Y., 2011a. Empirical study of risk assessment and allocation of public-private partnership projects in China. J. Manag. Eng. 27, 136–148.

Chan, D.W.M., Chan, A.P.C., Lam, P.T.I., Yeung, J.F.Y., Chan, J.H.L., 2011b. Risk ranking and analysis in target cost contracts: empirical evidence from the construction industry. Int. J. Proj. Manag. 29, 751–763.

Chan, A.P.C., Yi, W., Chan, D.W.M., Wong, D.P., 2013. Using the thermal work limit as an environmental determinant of heat stress for construction workers. J. Manag. Eng. 29, 414–423.

Chandramohan, A., Narayanan, S.L., Gaurav, A., Krishna, N., 2012. Cost and time overrun analysis for green construction projects. Int. J. Green Econ. 6, 167–177.

Chang, N.-B., Rivera, B.J., Wanielista, M.P., 2011. Optimal design for water conservation and energy savings using green roofs in a green building under mixed uncertainties. J. Clean. Prod. 19, 1180–1188.

Chanter, B., Swallow, P., 2007. Building Maintenance Management. Blackwell Publishing, Oxford, UK.

Chen, Y., Thomas Ng, S., James, M.W.W., 2012. Exploring the capability of Hong Kong beam plus for carbon footprint evaluation: perceptions of industry leaders. Int. J. Hous. Sci. Appl. 36, 211–219.

Chen, X., Yang, H., Lu, L., 2015. A comprehensive review on passive design approaches in green building rating tools. Renew. Sust. Energ. Rev. 50, 1425–1436.

Cheng, F., Li, H., Wang, Y.W., Skitmore, M., Forsythe, P., 2013. Modeling resource management in the building design process by information constraint petri nets. Autom. Constr. 29, 92–99.

Cheng, Y.-H., Lin, C.-C., Hsu, S.-C., 2015. Comparison of conventional and green building materials in respect of VOC emissions and ozone impact on secondary carbonyl emissions. Build. Environ. 87, 274–282.

Cherneva, I., 2012. The Business Case for Sustainable Finance. Taylor & Francis, New York, United States.

Chi, S., Han, S., 2013. Analyses of systems theory for construction accident prevention with specific reference to OSHA accident reports. Int. J. Proj. Manag. 31, 1027–1041.

Chi, C.-F., Chang, T.-C., Ting, H.-I., 2005. Accident patterns and prevention measures for fatal occupational falls in the construction industry. Appl. Ergon. 36, 391–400.

Chi, C.-F., Yang, C.-C., Chen, Z.-L., 2009. In-depth accident analysis of electrical fatalities in the construction industry. Int. J. Ind. Ergon. 39, 635–644.

Chi, C.F., Lin, Y.Y., Ikhwan, M., 2012. Flow diagram analysis of electrical fatalities in construction industry. Saf. Sci. 50, 1205–1214.

Chiang, Y.-H., Cheng Eddie, W.L., 2009. Perception of financial institutions toward financing PFI projects in Hong Kong. J. Constr. Eng. Manag. 135, 833–840.

Chien, K.F., Wu, Z.H., Huang, S.C., 2014. Identifying and assessing critical risk factors for bim projects: empirical study. Autom. Constr. 45, 1–15.

Choi, J., Peters, M., Mueller, R.O., 2010. Correlational analysis of ordinal data: from Pearson's R to Bayesian polychoric correlation. Asia Pac. Educ. Rev. 11, 459–466.

Choudhry, R., Fang, D., Ahmed, S., 2008. Safety management in construction: best practices in Hong Kong. J. Prof. Issues Eng. Educ. Pract. 134, 20–32.

Chua, G., 2015. South Australia Introduces Green Building Upgrade and Retrofit Financing. Available from: http://www.architectureanddesign.com.au/news/south-australia-introduces-green-building-upgrade (Accessed 20 June 2017).

Chua, D., Kog, Y., Loh, P., 1999. Critical success factors for different project objectives. J. Constr. Eng. Manag. 125, 142–150.

Chuing Loo, S., Abdul-Rahman, H., Wang, C., 2013. Managing external risks for international architectural, engineering, and construction (AEC) firms operating in Gulf Cooperation Council (GCC) states. Proj. Manag. J. 44, 70–88.

Clason, D.L., Dormody, T.J., 1994. Analyzing data measured by individual Likert-type items. J. Agric. Educ. 35, 4.

CMAA, 2012. Owners Guide to Project Delivery Methods. The Construction Management Association of America, USA.

Cole, R.J., Jose Valdebenito, M., 2013. The importation of building environmental certification systems: international usages of Breeam and Leed. Build. Res. Inf. 41, 662–676.

Comrey, A.L., 1973. A First Course in Factor Analysis. Academic Press, New York, NY.

Construction Industry Institute (CII), 2005. Making Zero Rework a Reality. The University of Texas at Austin, Austin, TX.

Construction Industry Institute (CII), 2008. Sustainable Design and Construction in Industrial Construction. Construction Industry Institute (CII), Austin, TX.

Construction Industry Institute (CII), 2010. Guide to Activity Analysis. Construction Industry Institute, United States.

Contreras, E., Buchanan, S., 2012. Piloting a personal protection equipment distribution program among Chicago day laborers. Am. J. Ind. Med. 55, 159–166.

Thal Jr., A.E., Cook, J.J., White III, E.D., 2010. Estimation of cost contingency for air force construction projects. J. Constr. Eng. Manag. 136, 1181–1188.

Cooke-Davies, T., 2002. The "real" success factors on projects. Int. J. Proj. Manag. 20, 185–190.

Creswell, J.W., 2013. Research Design: Qualitative, Quantitative, and Mixed Methods Approaches. Sage Publications, Thousand Oaks, United States.

Darko, A., Chan, A.P.C., 2016. Critical analysis of green building research trend in construction journals. Habitat Int. 57, 53–63.

Darko, A., Chan, A.P.C., 2017. Review of barriers to green building adoption. Sustain. Dev. 25, 167–179.

Darko, A., Zhang, C., Chan, A.P.C., 2017. Drivers for green building: a review of empirical studies. Habitat Int. 60, 34–49.

Davis Langdon Inc., 2007. Cost of Green Revisited: Reexamining the Feasibility and Cost Impact of Sustainable Design in the Light of Increased Market Adoption. Davis Langdon Inc., New York, NY.

De Marco, A., Mangano, G., 2013. Risk and value in privately financed health care projects. J. Constr. Eng. Manag. 139, 918–926.

De Winter, J.C.F., Dodou, D., 2010. Five-point Likert items: T test versus Mann-Whitney-Wilcoxon. Pract. Assess. Res. Eval. 15, 1–16.

Debrah, Y.A., Ofori, G., 1997. Flexibility, labour subcontracting and HRM in the construction industry in Singapore: can the system be refined? Int. J. Hum. Resour. Manag. 8, 690–709.

Deloitte, 2008. The Dollars and Sense of Green Retrofits. Deloitte Development LLC, New York, United States.

Demaid, A., Quintas, P., 2006. Knowledge across cultures in the construction industry: sustainability, innovation and design. Technovation 26, 603–610.

Demirkesen, S., Arditi, D., 2015. Construction safety personnel's perceptions of safety training practices. Int. J. Proj. Manag. 33, 1160–1169.

Deng, Y., Wu, J., 2014. Economic returns to residential green building investment: the developers' perspective. Reg. Sci. Urban Econ. 47, 35–44.

Deng, X., Low, S.P., Li, Q., Zhao, X., 2014a. Developing competitive advantages in political risk management for international construction enterprises. J. Constr. Eng. Manag. 140.

Deng, X., Pheng, L.S., Zhao, X., 2014b. Project system vulnerability to political risks in international construction projects: the case of Chinese contractors. Proj. Manag. J. 45, 20–33.

Department of Communities and Local Government, 2010. In: DOCAL Government (Ed.), Code for Sustainable Homes: Technical Guide November 2010. Eland House, Bressenden Place, London, UK.

Department of Energy, 2017. Database of State Incentives for Renewables & Efficiency. Available from: http://www.dsireusa.org/ (Accessed 16 July 2017).

Department of Statistics Singapore, 2014. Labour Productivity by Industry. Department of Statistics Singapore, Singapore. Available from: https://www.mti.gov.sg/researchroom/siteassets/pages/economic-survey-of-singapore-second-quarter-2014/a2.2_2q14.pdf (Accessed 4 April 2016).

Dewlaney, K.S., Hallowell, M., 2012. Prevention through design and construction safety management strategies for high performance sustainable building construction. Constr. Manag. Econ. 30, 165–177.

Dewlaney, K., Hallowell, M., Fortunato, B., 2011. Safety risk quantification for high performance sustainable building construction. J. Constr. Eng. Manag. 138, 964–971.

Dewlaney, K.S., Hallowell, M.R., Fortunato III, B.R., 2012. Safety risk quantification for high performance sustainable building construction. J. Constr. Eng. Manag. 138, 964–971.

Dikmen, I., Birgonul, M.T., Tah, J.H.M., Ozer, A.H., 2012. Web-based risk assessment tool using integrated duration-cost influence network model. J. Constr. Eng. Manag. 138, 1023–1034.

Ding, Z., Wang, Y., Zou, P.X.W., 2016. An agent based environmental impact assessment of building demolition waste management: conventional versus green management. J. Clean. Prod. 133, 1136–1153.

Doan, D.T., Ghaffarianhoseini, A., Naismith, N., Zhang, T.R., Ghaffarianhoseini, A., Tookey, J., 2017. A critical comparison of green building rating systems. Build. Environ. 123, 243–260.

Dodds, F., Strauss, M., Strong, M.F., 2012. Only One Earth: The Long Road Via Rio to Sustainable Development. Taylor & Francis, Abingdon, Oxford, United Kingdom.

Dodge Data & Analytics, 2016. World green building trends: developing markets accelerate global green growth smartmarket report. In: Design and Construction Intelligence. Dodge Data & Analytics, Research & Analytics, Bedford, MA.

Dogbegah, R., Owusu-Manu, D., Omoteso, K., 2011. A principal component analysis of project management competencies for the Ghanaian construction industry. Aust. J. Constr. Econ. Build. 11, 26–40.

Doloi, H., Sawhney, A., Iyer, K.C., Rentala, S., 2012. Analysing factors affecting delays in Indian construction projects. Int. J. Proj. Manag. 30, 479–489.

Dowson, M., Poole, A., Harrison, D., Susman, G., 2012. Domestic UK retrofit challenge: barriers, incentives and current performance leading into the green deal. Energy Policy 50, 294–305.

Drew, L., 2011. Careers in Green Construction. US Bureau of Labor Statistics, Washington, DC. Available from: http://www.bls.gov/green/construction/ (Accessed 11 November 2013).

Dulaimi, M.F., Dalziel, R.C., 1994. The Effects of the Procurement Method on the Level of Management Synergy in Construction Projects. CIB report, p. 53.

Durmus-Pedini, A., Ashuri, B., 2010. An overview of the benefits and risk factors of going green in existing buildings. Int. J. Facil. Manag. 1, 1–15.

Dwaikat, L.N., Ali, K.N., 2016. Green buildings cost premium: a review of empirical evidence. Energy Build. 110, 396–403.

Edum-Fotwe, F., McCaffer, R., 2000. Developing project management competency: perspectives from the construction industry. Int. J. Proj. Manag. 18, 111–124.

Edwards, P.J., Bowen, P.A., 1998. Risk and risk management in construction: a review and future directions for research. Eng. Constr. Archit. Manag. 5, 339–349.

Eisenberg, D., Done, R., Ishida, L., 2002. Breaking Down the Barriers: Challenges and Solutions to Code Approval of Green Building. Development Center for Appropriate Technology, Tucson, AZ.

Eisenstein, W., Fuertes, G., Kaam, S., Seigel, K., Arens, E., Mozingo, L., 2017. Climate co-benefits of green building standards: water, waste and transportation. Build. Res. Inf. 45, 828–844.

Ekambaram, P., Love, P.E.D., Kumaraswamy, M.M., Ng, T.S.T., 2014. Causal ascription of rework in building and civil engineering projects: a multivariate exploration. Eng. Constr. Archit. Manag. 21, 111–126.

El-Gohary, K.M., Aziz, R.F., 2014. Factors influencing construction labor productivity in Egypt. J. Manag. Eng. 30, 1–9.

El-Razek, M.E., Bassioni, H.A., Mobarak, A.M., 2008. Causes of delay in building construction projects in Egypt. J. Constr. Eng. Manag. ASCE 134 (11), 831–841.

El-Sayegh, S.M., 2008. Risk assessment and allocation in the UAE construction industry. Int. J. Proj. Manag. 26, 431–438.

El-Sayegh, S.M., Mansour, M.H., 2015. Risk assessment and allocation in highway construction projects in the UAE. J. Manag. Eng. 31, 04015004.

Environmental Protection Agency, 2016. Green Building Basic Information. Environmental Protection Agency, Singapore. Available from: http://www.epa.gov/greenbuilding/pubs/about.htm (Accessed 21 January 2016).

European Investment Bank, 2007. EPOS II—The "Climate Awareness Bond" EIB Promotes Climate Protection Via Pan-EU Public Offering. Available from: http://www.eib.org/eib.org/investor_relations/press/2007/2007-042-epos-ii-obligation-sensible-au-climat-la-bei-oeuvre-a-la-protection-du-climat-par-le-biais-de-son-emission-a-l-echelle-de-l-ue.htm?lang=en (Accessed 7 January 2017).

Eybpoosh, M., Dikmen, I., Birgonul, M.T., 2011. Identification of risk paths in international construction projects using structural equation modeling. J. Constr. Eng. Manag. 137, 1164–1175.

Fabian, N., 2015. Support low-carbon investment. Nature 519, 27.

Ferreira, J., Pinheiro, M.D., De Brito, J., 2014. Portuguese sustainable construction assessment tools benchmarked with Breeam and Leed: an energy analysis. Energy Build. 69, 451–463.

Finnegan, S., 2017. New Financial Strategies for Sustainable Buildings: Practical Guidance for Built Environment Professionals. Taylor & Francis Group, Abingdon, Oxford, United Kingdom.

Flyvbjerg, B., 2007. Policy and planning for large-infrastructure projects: problems, causes, cures. Environ. Plann. B Plann. Des. 34, 578–597.

Forcada, N., Rusiñol, G., Macarulla, M., Love, P.E.D., 2014. Rework in highway projects. J. Civ. Eng. Manag. 20, 455–465.

Ford, D.N., 2002. Achieving multiple project objectives through contingency management. J. Constr. Eng. Manag. 128, 30–39.

Fortunato III, B.R., Hallowell, M.R., Behm, M., Dewlaney, K., 2012. Identification of safety risks for high-performance sustainable construction projects. J. Constr. Eng. Manag. 138, 499–508.

Fortunato, B.R., Hallowell, M.R., Behm, M., Dewlaney, K., 2011. Identification of safety risks for high-performance sustainable construction projects. J. Constr. Eng. Manag. 138, 499–508.

Fortunato, B.R., Hallowell, M.R., Behm, M., Dewlaney, K., 2012. Identification of safety risks for high-performance sustainable construction projects. J. Constr. Eng. Manag. ASCE 138, 499–508.

Fowler, K.M., Rauch, E.M., 2008. Assessing Green Building Performance: A Post Occupancy Evaluation of 12 GSA Buildings. Pacific Northwest National Laboratory Report Number PNNL-17393.

Fox, S., Marsh, L., Cockerham, G., 2002. How building design imperatives constrain construction productivity and quality. Eng. Constr. Archit. Manag. 9, 378–387.

Frame, J.D., 2003. Managing Projects in Organizations. Jossey-Bass, San Francisco, CA.

Frank, T., 2001. The Superior Project Manager: Global Competency Standards and Best Practices. CRC Press, Boca Raton, FL, United States.

Frank, T., 2002. The Superior Project Manager. Marcel Dekker, New York, NY.

Fraser, C., 1999. A non-results-based effectiveness index for construction site managers. Constr. Manag. Econ. 17, 789–798.

Fu, T., Chen, H.Z., Choi, Y., 2016. Does the sustainable PPI investments promote financial market's sustainable development? Sustainability 8, 120.

Furr, J.E., American Bar Association, Section of Real Property, Trust & Estate Law, 2009. Green Building and Sustainable Development: The Practical Legal Guide. Section of Real Property, Trust and Estate Law, American Bar Association.

Ganah, A., Henderson, C., John, G., 2015. Feasibility of CSH in assisting the new housing sector in delivering zero carbon homes By 2016. Procedia Eng. 118, 1000–1007.

Gangolells, M., Casals, M., Forcada, N., Roca, X., Fuertes, A., 2010. Mitigating construction safety risks using prevention through design. J. Saf. Res. 41, 107–122.

Geng, Y., Dong, H., Xue, B., Fu, J., 2012. An overview of Chinese green building standards. Sustain. Dev. 20, 211–221.

Geoghegan, L., Dulewicz, V., 2008. Do project managers' leadership competencies contribute to project success? Proj. Manag. J. 39, 58–67.

Ghaffarianhoseini, A., Dahlan, N.D., Berardi, U., Ghaffarianhoseini, A., Makaremi, N., Ghaffarianhoseini, M., 2013. Sustainable energy performances of green buildings: a review of current theories, implementations and challenges. Renew. Sust. Energ. Rev. 25, 1–17.

Glavinich, T.E., 2008. Contractor's Guide To Green Building Construction: Management, Project Delivery. John Wiley & Sons, Hoboken, NJ.

Global Environment Facility, 2016. From Efficient Buildings to Sustainable Cities in 8 Action Steps. Available from: https://www.thegef.org/news/efficient-buildings-sustainable-cities-8-action-steps (Accessed 7 November 2017).

Global Environment Facility, 2017. Funding. Available from: https://www.thegef.org/about/funding# (Accessed 7 October 2017).

Gluch, P., Gustafsson, M., Thuvander, L., 2009. An absorptive capacity model for green innovation and performance in the construction industry. Constr. Manag. Econ. 27, 451–464.

Goh, B.H., 2016. Designing a whole-life building cost index in Singapore. Built Environ. Project Asset Manag. 6, 159–173.

Goh, Y.M., Chua, D., 2013. Neural network analysis of construction safety management systems: a case study in Singapore. Constr. Manag. Econ. 31, 460–470.

Goodrum, P.M., Haas, C.T., 2002. Partial factor productivity and equipment technology change at activity level in U.S. construction industry. J. Constr. Eng. Manag. 128, 463–472.

Goodrum, P.M., Haas, C.T., 2004. Long-term impact of equipment technology on labor productivity in the U.S. construction industry at the activity level. J. Constr. Eng. Manag. 130, 124–133.

Gou, Z., Lau, S.S.Y., 2014. Contextualizing green building rating systems: case study of Hong Kong. Habitat Int. 44, 282–289.

Gowri, K., 2004. Green building rating systems: an overview. ASHRAE J. 46, 56.

Gravetter, F., Forzano, L.-A., 2012. Research Methods for the Behavioral Sciences. Cengage Learning, Belmont, CA.

Green Building Council, 2016. Green Star Technical Manual 3.1. Available from: https://www.nzgbc.org.nz/attachment?action=download&attachment_id=694 (Accessed 1 October 2017).

Green Building Council, 2017. Green Finance. Available from: http://www.ukgbc.org/resources/key-topics/cities-and-communities/green-finance (Accessed 14 June 2017).

Green Building Council Australia, 2008. The Dollars and Sense of Green Buildings 2008. Australia.

Green Building Index, 2013. Green Building Index Assessment Criteria for Residential New Construction. Available from: http://new.greenbuildingindex.org/files/resources/gbi%20tools/gbi%20rnc%20residential%20tool%20v3.0.pdf (Accessed 10 April 2017).

Green Building Initiative, 2017. Green Globes for New Construction. Available from: https://www.thegbi.org/green-globes-certification/how-to-certify/new-construction/ (Accessed 10 April 2017).

Green Investment Bank, 2017. Investing to Build a Stronger, Greener UK Economy. Available from: http://www.greeninvestmentbank.com/what-we-do/ (Accessed 14 June 2017).

Greenbiz, 2005. Green Building Technique: A Two-Minute Briefing on Key Business Environmental Issues. Greenbiz Group Inc., Oakland, CA.

Greenwald, J., 2012. Green buildings pose myriad risks. Bus. Insur. 46, 03.

Griffith, A., 2002. Management systems for sustainable construction: integrating environmental, quality and safety management systems. Int. J. Environ. Technol. Manag. 2, 114–126.

Group Twenty, 2016. G20 Action Plan on the 2030 Agenda for Sustainable Development. Hangzhou, China. Available from: https://www.g20.org/content/de/_anlagen/g7_g20/2016-09-08-g20-agenda-action-plan.pdf?__blob=publicationfile&v=4 (Accessed 18 September 2017).

Gruneberg, S., Hughes, W., Ancell, D., 2007. Risk under performance-based contracting in the UK construction sector. Constr. Manag. Econ. 25, 691–699.

Gündüz, M., Nielsen, Y., Özdemir, M., 2013. Quantification of delay factors using the relative importance index method for construction projects in Turkey. J. Manag. Eng. 29, 133–139.

Guo, B.H.W., Yiu, T.W., 2016. Developing leading indicators to monitor the safety conditions of construction projects. J. Manag. Eng. 32, 0000376.

Gushgari, S.K., Francis, P.A., Saklou, J.H., 1997. Skills critical to long-term profitability of engineering firms. J. Manag. Eng. 13, 46–56.

Haapio, A., Viitaniemi, P., 2008. A critical review of building environmental assessment tools. Environ. Impact Assess. Rev. 28, 469–482.

Hajmohammad, S., Vachon, S., 2014. Safety culture: a catalyst for sustainable development. J. Bus. Ethics 123, 263–281.

Häkkinen, T., Belloni, K., 2011. Barriers and drivers for sustainable building. Build. Res. Inf. 39, 239–255.

Hallowell, M.R., Gambatese, J.A., 2009. Construction safety risk mitigation. J. Constr. Eng. Manag. 135, 1316–1323.

Halper, M., 2015. Green Investment Bank Launches Industry-Wide Led Retrofit Funding. First Stop, GKN Aerospace. Available from: http://luxreview.com/article/2015/06/green-investment-bank-launches-industry-wide-led-retrofit-scheme-at-gkn-aerospace (Accessed 14 June 2017).

Hamilton, K., 2007. National Association of Industrial and Office Properties Gives Green Awards. Available from: http://www.igreenbuild.com/cd_2940.aspx (Accessed 13 July 2017).

Han, S.H., Park, S.H., Jin, E.J., Kim, H., Seong, Y.K., 2008. Critical issues and possible solutions for motivating foreign construction workers. J. Manag. Eng. 24, 217–226.

Haslam, R.A., Hide, S.A., Gibb, A.G., Gyi, D.E., Pavitt, T., Atkinson, S., Duff, A., 2005. Contributing factors in construction accidents. Appl. Ergon. 36, 401–415.

Hawksford, S., 2009. Major Industries in Singapore. Available from: http://www.guidemesingapore.com/blog-post/singapore-life/major-industries-in-singapore (Accessed 20 April 2016).

He, Q., Wang, G., Luo, L., Shi, Q., Xie, J., Meng, X., 2017. Mapping the managerial areas of building information modeling (BIM) using scientometric analysis. Int. J. Proj. Manag. 35, 670–685.

Hersey, P., Blanchard, K.H., 1974. So you want to know your leadership style? Train. Dev. J. 28, 22–37.

Hersey, P., Blanchard, K.H., 1993. Management of Organizational Behavior: Utilizing Human Resources. Prentice Hall, Englewood Cliffs, NJ.

Hewage, K.N., Ruwanpura, J.Y., 2006. Carpentry workers issues and efficiencies related to construction productivity in commercial construction projects in Alberta. Can. J. Civ. Eng. 33, 1075–1089.

Hewage, K.N., Ruwanpura, J.Y., 2009. A novel solution for construction on-site communication—the information booth. Can. J. Civ. Eng. 36, 659–671.

Hill, G., 1975. U.N. environment effort: a start, a long way to go. Available from: http://www.nytimes.com/1975/10/20/archives/un-environment-effort-a-start-a-long-way-to-go-un.html?_r=0 (Accessed 15 June 2017).

Hinze, J., 1997. Construction Safety. Prentice Hall, Upper Saddle River, NJ, United States.

Hinze, J.W., Teizer, J., 2011. Visibility-related fatalities related to construction equipment. Saf. Sci. 49, 709–718.

Hlaing, N.N., Singh, D., Tiong, R.L.K., Ehrlich, M., 2008. Perceptions of singapore construction contractors on construction risk identification. J. Financ. Manag. Prop. Constr. 13, 85–95.

Hoffman, A.J., Henn, R., 2008. Overcoming the social and psychological barriers to green building. Org. Environ. 21, 390–419.

Hong Kong Financial Services Development Council, 2016. Hong Kong as a Regional Green Finance Hub. Hong Kong.

Horn, J.L., 1965. A rationale and test for the number of factors in factor analysis. Psychometrika 30, 179–185.

Houghton, A., Vittori, G., Guenther, R., 2009. Demystifying first-cost green building premiums in healthcare. HERD 2, 10–45.

House, R.J., 1971. A path goal theory of leader effectiveness. Adm. Sci. Q. 16, 321–339.

House, R.J., 1996. Path goal theory of leadership: lessons, legacy and a reformulated theory. Leadersh. Q. 7, 323–352.

Howe, J.C., Gerrard, M., Fucci, F.R., American Bar Association, Section of Environment, Energy & Resources, 2010. The Law of Green Buildings: Regulatory and Legal Issues in Design, Construction, Operations, and Financing. American Bar Association, Section of Environment, Energy, and Resources.

Huang, X., Hinze, J., 2003. Analysis of construction worker fall accidents. J. Constr. Eng. Manag. 129, 262–271.

Huo, X., Yu, A.T.W., Wu, Z., 2017. A comparative analysis of site planning and design among green building rating tools. J. Clean. Prod. 147, 352–359.

Hwang, B.G., Leong, L.P., 2013. Comparison of schedule delay and causal factors between traditional and green construction projects. Technol. Econ. Dev. Econ. 19, 310–330.

Hwang, B.-G., Ng, W.J., 2013. Project management knowledge and skills for green construction: overcoming challenges. Int. J. Proj. Manag. 31, 272–284.

Hwang, B.G., Tan, J.S., 2012. Green building project management: obstacles and solutions for sustainable development. Sustain. Dev. 20, 335–349.

Hwang, B.G., Yang, S., 2014. Rework and schedule performance: a profile of incidence, impact, causes and solutions. Eng. Constr. Archit. Manag. 21, 190–205.

Hwang, B.G., Thomas, S.R., Haas, C.T., Caldas, C.H., 2009. Measuring the impact of rework on construction cost performance. J. Constr. Eng. Manag. 135, 187–198.

Hwang, B.-G., Leong, L.P., Huh, Y.-K., 2013a. Sustainable green construction management: schedule performance and improvement. Technol. Econ. Dev. Econ. 19, S43–S57.

Hwang, B., Zhao, X., Gay, M.J.S., 2013b. Public private partnership projects in Singapore: factors, critical risks and preferred risk allocation from the perspective of contractors. Int. J. Proj. Manag. 31, 424–433.

Hwang, B.G., Zhao, X., Ng, S.Y., 2013c. Identifying the critical factors affecting schedule performance of public housing projects. Habitat Int. 38, 214–221.

Hwang, B.G., Zhao, X., Goh, K.J., 2014. Investigating the client-related rework in building projects: the case of Singapore. Int. J. Proj. Manag. 32, 698–708.

Hwang, B.-G., Zhao, X., See, Y.L., Zhong, Y., 2015a. Addressing risks in green retrofit projects: the case of Singapore. Proj. Manag. J. 46, 76–89.

Hwang, B.-G., Zhao, X., Tan, L.L.G., 2015b. Green building projects: schedule performance, influential factors and solutions. Eng. Constr. Archit. Manag. 22.

Hwang, B.G., Zhao, X., Ong, S., 2015c. Value management in Singaporean building projects: implementation status, critical success factors, and risk factors. J. Manag. Eng. 31.

Hwang, B.G., Zhao, X., Tan, L.L.G., 2015d. Green building projects: schedule performance, influential factors and solutions. Eng. Constr. Archit. Manag. 22, 327–346.

Hwang, B.G., Zhao, X., Yu, G.S., 2015e. Risk identification and allocation in underground rail construction joint ventures: contractors' perspective. J. Civ. Eng. Manag. https://doi.org/10.3846/13923730.2014.914095.

Hwang, B., Shan, M., Tan, E., 2016. Investigating reworks in green building construction projects: magnitude, influential factors, and solutions. Int. J. Environ. Res. 10, 499–510.

Hwang, B.-G., Shan, M., Phua, H., Chi, S., 2017a. An exploratory analysis of risks in green residential building construction projects: the case of Singapore. Sustainability 9, 1116.

Hwang, B.-G., Shan, M., Phuah, S.L., 2017b. Safety in green building construction projects in Singapore: performance, critical issues, and improvement solutions. KSCE J. Civ. Eng., 1–12.

Hwang, B.-G., Shan, M., Supa'at, N.N.B., 2017c. Green commercial building projects in Singapore: critical risk factors and mitigation measures. Sustain. Cities Soc. 30, 237–247.

Hwang, B.-G., Shan, M., Xie, S., Chi, S., 2017d. Investigating residents' perceptions of green retrofit program in mature residential estates: the case of Singapore. Habitat Int. 63, 103–112.

Hydes, K.R., Creech, L., 2000. Reducing mechanical equipment cost: the economics of green design. Build. Res. Inf. 28 (5), 403–407.

ICF International, 2007. Introduction to Energy Performance Contracting. U.S. Environmental Protection Agency, Energy Star Buildings.

IFC, 2016. Green Bond Impact Report-Financial Year 2016. International Finance Corporation.

IFC, 2017a. About IFC—Overview. Available from: http://www.ifc.org/wps/wcm/connect/corp_ext_content/ifc_external_corporate_site/about+ifc_new/ (Accessed 19 June 2017).

IFC, 2017b. Overview of IFC's Green Bonds. Available from: http://www.ifc.org/wps/wcm/connect/corp_ext_content/ifc_external_corporate_site/about+ifc_new/ifc+governance/investor+relations/grnbond-overvw (Accessed 19 June 2017).

Illankoon, I.M.C.S., Tam, V.W.Y., Le, K.N., 2017a. Environmental, economic, and social parameters in international green building rating tools. J. Prof. Issues Eng. Educ. Pract. 143, 05016010.

Illankoon, I.M.C.S., Tam, V.W.Y., Le, K.N., Shen, L., 2017b. Key credit criteria among international green building rating tools. J. Clean. Prod. 164, 209–220.

Ilvonen, O., 2013. Assessing release of hazardous substances from construction products—review of 10 years of experience with a horizontal approach in the European Union. Build. Environ. 69, 194–205.

Indian Green Building Council, 2014. IGBC Green New Buildings Rating System Version 3.0. Available from: https://igbc.in/igbc/html_pdfs/abridged/igbc%20green%20new%20buildings%20rating%20system%20(version%203.0).pdf (Accessed 10 April 2017).

Institute for Building Environment and Energy Conservation, 2014. Comprehensive Assessment System for Built Environment Efficiency: Technical Manual 2014 Edition. Available from: http://www.ibec.or.jp/casbee/english/download/casbee-bd(nc)e_2014manual.pdf (Accessed 10 March 2017).

Ireland, V., 1985. The role of managerial actions in the cost, time and quality performance of high-rise commercial building projects. Constr. Manag. Econ. 3, 59–87.

Irumba, R., 2014. Spatial analysis of construction accidents in Kampala, Uganda. Saf. Sci. 64, 109–120.

Ismail, S., 2013. 3 New Requirements for Buildings to be Energy-Efficient. Channel New Asia. Available from: http://www.channelnewsasia.com/news/singapore/3-new-requirements-for-bu/482052.html (Accessed 24 August 2016).

Iyer, K.C., Jha, K.N., 2006. Critical factors affecting schedule performance: evidence from indian construction projects. J. Constr. Eng. Manag. 132, 871–881.

Iyer, K.C., Sagheer, M., 2010. Hierarchical structuring of PPP risks using interpretative structural modeling. J. Constr. Eng. Manag. 136, 151–159.

Jarkas, A.M., Bitar, C.G., 2014. Factors affecting construction labor productivity in Kuwait. J. Constr. Eng. Manag. 138, 811–820.

Jaselskis, E.J., Recarte Suazo, G.A., 1994. A survey of construction site safety in Honduras. Constr. Manag. Econ. 12, 245–255.

Jeucken, M., 2010. Sustainable Finance and Banking: The Financial Sector and the Future of the Planet. Taylor & Francis, Abingdon, Oxford, United Kingdom.

Jha, K.N., Devaya, M.N., 2008. Modelling the risks faced by Indian construction companies assessing international projects. Constr. Manag. Econ. 26, 337–348.

Jiang, Y., 2010. Understanding the Cost of Green Buildings: Evidence From Singapore. Master's thesis, National University of Singapore.

Jin, X.H., 2010. Determinants of efficient risk allocation in privately financed public infrastructure projects in Australia. J. Constr. Eng. Manag. 136, 138–150.

Johnson, S.D., 2000. The economic case for "high performance buildings". Corp. Environ. Strateg. 7, 350–361.

Josephson, P.E., Larsson, B., Li, H., 2002. Illustrative benchmarking rework and rework costs in Swedish construction industry. J. Manag. Eng. 18, 76–83.

Kaming, P.F., Olomolaiye, P.O., Holt, G.D., Harris, F.C., 1997. Factors influencing craftsmen's productivity in Indonesia. Int. J. Proj. Manag. 15, 21–30.

Kang, Y., Kim, C., Son, H., Lee, S., Limsawasd, C., 2013. Comparison of preproject planning for green and conventional buildings. J. Constr. Eng. Manag. 139.

Kansal, R., Kadambari, G., 2010. Green buildings: an assessment of life cycle cost. IUP J. Infrastruct. VIII (4), 50–57.

Karakhan, A.A., 2016. Leed-certified projects: green or sustainable? J. Manag. Eng. 32, 02516001.

Kartam, N.A., Kartam, S.A., 2001. Risk and its management in the Kuwaiti construction industry: a contractors' perspective. Int. J. Proj. Manag. 19, 325–335.

Kaskutas, V., Dale, A.M., Lipscomb, H., Evanoff, B., 2013. Fall prevention and safety communication training for foremen: report of a pilot project designed to improve residential construction safety. J. Saf. Res. 44, 111–118.

Kats, G., 2010. Greening Our Built World: Costs, Benefits, and Strategies. Island Press, Washington, DC.

Kats, G., Alevantis, L., Berman, A., Mills, E., Perlman, J., 2003. The Costs and Financial Benefits of Green Buildings: A Report to California's Sustainable Building Task Force. Capital E, California, USA.

Ke, Y., Wang, S., Chan, A.P., Cheung, E., 2011. Understanding the risks in China's PPP projects: ranking of their probability and consequence. Eng. Constr. Archit. Manag. 18, 481–496.

Kerzner, H., 2013. Project Management: A Systems Approach to Planning, Scheduling, and Controlling. John Wiley & Sons, New York, United States.

Khodakarami, V., Abdi, A., 2014. Project cost risk analysis: a Bayesian networks approach for modeling dependencies between cost items. Int. J. Proj. Manag. 32, 1233–1245.

Kiat, K.J., 1991. Productivity 2000: Singapore's initiative to maintain competitiveness. Work Study 40, 15–17.

Kibert, C.J., 2008. Sustainable Construction: Green Building Design and Delivery. John Wiley & Sons, Hoboken, NJ.

Kibert, C.J., 2012. Sustainable Construction: Green Building Design and Delivery. John Wiley & Sons, Hoboken, NJ.

Kibert, C.J., 2016. Sustainable Construction: Green Building Design and Delivery. John Wiley & Sons, New York, United States.

Kim Cheng Low, P., 2006. Father leadership: the Singapore case study. Manag. Decis. 44, 89–104.

Kim, S., Osmond, P., 2014. Analyzing green building rating tools for healthcare buildings from the building user's perspective. Indoor Built Environ. 23, 757–766.

Kim, J.L., Greene, M., Kim, S., 2014. Cost comparative analysis of a new green building code for residential project development. J. Constr. Eng. Manag. 140 (5), 05014002.

Kines, P., 2003. Case studies of occupational falls from heights: cognition and behavior in context. J. Saf. Res. 34, 263–271.

Kirkpatrick, S.A., Locke, E.A., 1991. Leadership: do traits mater? Acad. Manag. Exec. 5, 48–60.

Koksal, A., Arditi, D., 2004. Predicting construction company decline. J. Constr. Eng. Manag. 130, 799–807.

Korkmaz, S., Messner, J.I., Riley, D.R., Magent, C., 2010. High-performance green building design process modeling and integrated use of visualization tools. J. Archit. Eng. 16, 37–45.

Kothari, C.R., 2004. Research Methodology: Methods and Techniques. New Age International, New Delhi, India.

Kubba, S., 2010. Green Construction Project Management and Cost Oversight. Butterworth-Heinemann, London, United Kingdom.

Kubba, S., 2012. Handbook of Green Building Design and Construction: Leed, Breeam, and Green Globes. Butterworth-Heinemann, Waltham.

Lam, K.C., Chow, W.S., 1999. The significance of financial risks in BOT procurement. Build. Res. Inf. 27, 84–95.

Lam, P.T.I., Chan, E.H.W., Chau, C.K., Poon, C.S., Chun, K.P., 2009. Integrating green specifications in construction and overcoming barriers in their use. J. Prof. Issues Eng. Educ. Pract. 135, 142–152.

Lam, P.T.I., Chan, E.H.W., Poon, C.S., Chau, C.K., Chun, K.P., 2010. Factors affecting the implementation of green specifications in construction. J. Environ. Manag. 91, 654–661.

Lam, P.T., Chan, E.H., Chau, C., Poon, C., Chun, K., 2011. Environmental management system vs green specifications: how do they complement each other in the construction industry? J. Environ. Manag. 92, 788–795.

Langdon, D., 2007. The Cost & Benefit of Achieving Green Buildings. Davis Langdon Management Consulting, Australia.

Lapinski, A.R., Horman, M.J., Riley, D.R., 2006. Lean processes for sustainable project delivery. J. Constr. Eng. Manag. 132, 1083–1091.

Larsson, N., 2016. Overview of the SBTool Assessment Framework. Available from: http://www.iisbe.org/system/files/private/sbtool%202016%20description%2021jul16.pdf (Accessed 1 October 2017).

Le, T., Caldas, C.H., Gibson, G.E., Thole, M., 2009. Assessing scope and managing risk in the highway project development process. J. Constr. Eng. Manag. 135, 900–910.

Le, Y., Shan, M., Chan, A.P.C., Hu, Y., 2014. Overview of corruption research in construction. J. Manag. Eng. 30.

Lee, W.L., Yik, F.W.H., Burnett, J., 2007. Assessing energy performance in the latest versions of Hong Kong Building Environmental Assessment Method (HK-BEAM). Energy Build. 39, 343–354.

Lee, J., Edil, T.B., Benson, C.H., Tinjum, J.M., 2013a. Building environmentally and economically sustainable transportation infrastructure: green highway rating system. J. Constr. Eng. Manag. 139, 10.

Lee, S., Lee, B., Kim, J., Kim, J., 2013b. A financing model to solve financial barriers for implementing green building projects. Sci. World J. 2013, 240394.

Lee, H.W., Tommelein, I.D., Ballard, G., 2015. Target-setting practice for loansfor commercial energy-retrofit projects. J. Manag. Eng. 31.

Lee-Kelley, L., Leong, L., 2003. Turner's five-functions of project-based management and situational leadership in it services projects. Int. J. Proj. Manag. 21, 583–591.

Lewin, K., Lippitt, R., White, R.K., 1939. Patterns of aggressive behavior in experimentally created "social climates" J. Soc. Psychol. 10, 269–299.

Li, B., Akintoye, A., Corresponding, P.J.E., Hardcastle, C., 2005. Critical success factors for PPP/PFI projects in the UK construction industry. Constr. Manag. Econ. 23, 459–471.

Li, Y.Y., Chen, P.H., Chew, D.A.S., Teo, C.C., Ding, R.G., 2011. Critical project management factors of AEC firms for delivering green building projects in Singapore. J. Constr. Eng. Manag. 137, 1153–1163.

Li, Y., Yang, L., He, B.-J., Zhao, D., 2013. Green building in China: needs great promotion. Sustain. Cities Soc. 11, 1–6.

Li, Y., Chen, X., Wang, X., Xu, Y., Chen, P.H., 2017. A review of studies on green building assessment methods by comparative analysis. Energy Build. 146, 152–159.

Liberty Property Trust, 2014. Sustainability. Available from: https://www.libertyproperty.com/sustainability.asp (Accessed 1 October 2017).

Likert, R., 1961. New Patterns of Management. McGraw-Hill, New York, NY.

Lim, E.C., Alum, J., 1995. Construction productivity: issues encountered by contractors in Singapore. Int. J. Proj. Manag. 13, 51–58.

Lim, Y.S., Xia, B., Skitmore, M., Gray, J., Bridge, A., 2015. Education for sustainability in construction management curricula. Int. J. Constr. Manag. 15, 321–331.

Ling, J.U., 2003. The project manager's personal characteristic, skills and roles in local construction industry. Published Master's dissertation, Faculty of Civil Engineering, University Technology Malaysia.

Ling, F.Y.Y., Hoang, V.T.P., 2009. Political, economic, and legal risks faced in international projects: case study of Vietnam. J. Prof. Issues Eng. Educ. Pract. 136, 156–164.

Ling, F.Y.Y., Hoi, L., 2006. Risks faced by Singapore firms when undertaking construction projects in India. Int. J. Proj. Manag. 24, 261–270.

Ling, F.Y.Y., Tiong, J.H., 2008. Managing large commercial projects with a tight schedule in China: a case study. Int. J. Project Org. Manag. 1, 86–104.

Ling, F.Y.Y., Low, S.P., Wang, S.Q., Lim, H.H., 2009. Key project management practices affecting Singaporean firms' project performance in China. Int. J. Proj. Manag. 27, 59–71.

Lingard, H., Rowlinson, S.M., 2005. Occupational Health and Safety in Construction Project Management. Taylor & Francis, London, United Kingdom.

Lipscomb, H.J., Schoenfisch, A.L., Shishlov, K.S., 2010. Non-fatal contact injuries among workers in the construction industry treated in US emergency departments, 1998–2005. J. Saf. Res. 41, 191–195.

Liu, M., Ballard, G., Ibbs, W., 2011. Work flow variation and labor productivity: case study. J. Manag. Eng. 27, 236–242.

Liu, J.Y., Low, S.P., He, X., 2012. Green practices in the Chinese building industry: drivers and impediments. J. Technol. Manag. China 7, 50–63.

Liu, J., Li, Q., Wang, Y., 2013. Risk analysis in ultra deep scientific drilling project—a fuzzy synthetic evaluation approach. Int. J. Proj. Manag. 31, 449–458.

Liu, J., Zhao, X., Li, Y., 2016. Exploring the factors inducing contractors' unethical behavior: case of China. J. Prof. Issues Eng. Educ. Pract.

Lockwood, C., 2009. Building retrofits. Urban Land, 46–57.

Long, N.D., Ogunlana, S., Quang, T., Lam, K.C., 2004. Large construction projects in developing countries: a case study from Vietnam. Int. J. Proj. Manag. 22, 553–561.

Long, L.-H., Lee, Y.D., Lee, J.Y., 2008. Delay and cost overruns in Vietnam large construction projects: a comparison with other selected countries. KSCE J. Civ. Eng. 12, 367–377.

Love, P.E.D., 2002. Influence of project type and procurement method on rework costs in building construction projects. J. Constr. Eng. Manag. 128, 18–29.

Love, P.E.D., Edwards, D.J., 2004. Determinants of rework in building construction projects. Eng. Constr. Archit. Manag. 11, 259–274.

Love, P.E.D., Li, H., 2000. Quantifying the causes and costs of rework in construction. Constr. Manag. Econ. 18, 479–490.

Love, P.E.D., Smith, J., 2003. Benchmarking, benchaction, and benchlearning: rework mitigation in projects. J. Manag. Eng. 19, 147–159.

Love, P.E., Mandal, P., Li, H., 1999. Determining the causal structure of rework influences in construction. Constr. Manag. Econ. 17, 505–517.

Love, P.E.D., Mandal, P., Smith, J., Heng, L.I., 2000. Modelling the dynamics of design error induced rework in construction. Constr. Manag. Econ. 18, 567–574.

Love, P., Edwards, D., Watson, H., Davis, P., 2010. Rework in civil infrastructure projects: determination of cost predictors. J. Constr. Eng. Manag. 136, 275–282.

Love, P.E.D., Edwards, D.J., Irani, Z., Goh, Y.M., 2011. Dynamics of rework in complex offshore hydrocarbon projects. J. Constr. Eng. Manag. 137, 1060–1070.

Love, P.E.D., Niedzweicki, M., Bullen, P.A., Edwards, D.J., 2012. Achieving the green building council of Australia's world leadership rating in an office building in Perth. J. Constr. Eng. Manag. 138, 652–660.

Love, P.E.D., Edwards, D.J., Irani, Z., Forcada, N., 2014. The latent causes of rework in floating production storage and offloading projects. J. Civ. Eng. Manag. 20, 315–329.

Love, P.E., Liu, J., Matthews, J., Sing, C.-P., Smith, J., 2015a. Future proofing PPPs: life-cycle performance measurement and building information modelling. Autom. Constr. 56, 26–35.

Love, P.E.D., Ackermann, F., Teo, P., Morrison, J., 2015b. From individual to collective learning: a conceptual learning framework for enacting rework prevention. J. Constr. Eng. Manag. 141 (11), 05015009.

Low, S.P., 2001. Quantifying the relationships between buildability, structural quality and productivity in construction. Struct. Surv. 19, 106–112.

Low, L., 2002. The political economy of migrant worker policy in Singapore. Asia Pac. Bus. Rev. 8, 95–118.

Low, S.P., Shi, Y., 2002. An exploratory study of Hofstede's cross-cultural dimensions in construction projects. Manag. Decis. 40, 7–16.

Lu, S., Yan, H., 2013. A comparative study of the measurements of perceived risk among contractors in China. Int. J. Proj. Manag. 31, 307–312.

Lu, W., Yuan, H., 2013. Investigating waste reduction potential in the upstream processes of offshore prefabrication construction. Renew. Sust. Energ. Rev. 28, 804–811.

Lupsea, M., Tiruta-Barna, L., Schiopu, N., 2014. Leaching of hazardous substances from a composite construction product—an experimental and modelling approach for fibre-cement sheets. J. Hazard. Mater. 264, 236–245.

Lutzkendorf, T., Fan, W., Lorenz, D., 2011. Engaging financial stakeholders: opportunities for a sustainable built environment. Build. Res. Inf. 39, 483–503.

Ma, J., Kremer, G.E.O., 2015. A fuzzy logic-based approach to determine product component end-of-life option from the views of sustainability and designer's perception. J. Clean. Prod. https://doi.org/10.1016/J.Jclepro.2015.08.029.

Maccallum, R.C., Widaman, K.F., Zhang, S., Hong, S., 1999. Sample size in factor analysis. Psychol. Methods 4, 84–99.

Magent, C.S., Korkmaz, S., Klotz, L.E., Riley, D.R., 2009. A design process evaluation method for sustainable buildings. Archit. Eng. Des. Manag. 5, 62–74.

Mahamid, I., 2013. Contractors perspective toward factors affecting labor productivity in building construction. Eng. Constr. Archit. Manag. 20, 446–460.

Majld, M.Z., McCaffer, R., 1998. Factors of non-excuseable delays that influence contractors' performance. J. Manag. Eng. ASCE 14 (3), 42–49.

Makulsawatudom, A., Emsley, M., Sinthawanarong, K., 2004. Critical factors influencing construction productivity in Thailand. J. KMITNB 14, 1–6.

Male, S., 2003. Faster building for commerce: NEDO (1988). In: Murray, M., Langford, D. (Eds.), Construction Reports 1944–98. Blackwell Science Ltd, Oxford, UK.

Malhotra, N.K., 2006. Marketing Research: An Applied Orientation. Pearson/Prentice Hall, Upper Saddle River, NJ.

Malin, N., 2000. The cost of green materials. Build. Res. Inf. 28, 408–412.

Malina, M., 2012. Delivering Sustainable Buildings: An Industry Insider's View. Wiley, New York, United States.

Mansfield, N.R., Ugwu, O.O., Doran, T., 1994. Causes of delay and cost overruns in Nigerian construction projects. Int. J. Project Manag. 12 (4), 254–260.

Mao, X., Lu, H., Li, Q., 2009. A comparison study of mainstream sustainable/green building rating tools in the world. In: International Conference on Management and Service Science, 2009, Mass'09, IEEE.pp. 1–5.

Marcelo, C., Daniel, P., 2013. Green performs better: energy efficiency and financial return on buildings. J. Corp. Real Estate 15, 53–72.

Marques, R.C., Berg, S., 2011. Risks, contracts, and private-sector participation in infrastructure. J. Constr. Eng. Manag. 137, 925–932.

Matsunaga, M., 2010. How to factor-analyze your data right: do's, don'ts, and how-to's. Int. J. Psychol. Res. 3, 97–110.

Mattoni, B., Guattari, C., Evangelisti, L., Bisegna, F., Gori, P., Asdrubali, F., 2018. Critical review and methodological approach to evaluate the differences among international green building rating tools. Renew. Sust. Energ. Rev. 82, 950–960.

Maxwell, J.A., 2012. Qualitative Research Design: An Interactive Approach. SAGE Publications, Newcastle Upon Tyne, United Kingdom.

Mccann, M., 2006. Heavy equipment and truck-related deaths on excavation work sites. J. Saf. Res. 37, 511–517.

McGraw Hill Construction, 2012. Construction Industry Workforce Shortages: Role of Certification, Training and Green Jobs in Filling the Gaps. SmartMarket report.

McGraw-Hill Construction, 2006a. Green Building Smart Market Report. McGraw-Hill Construction, Lexington, MA.

McGraw-Hill Construction, 2006b. Green Building Smart Market Report: Design & Construction Intelligence. McGraw-Hill, New York, NY.

McGraw-Hill Construction, 2013. World Green Building Trends: Business Benefits Driving New and Retrofit Market Opportunities in Over 60 Countries. McGraw-Hill Construction, Bedford, MA.

Mcpherson, D., 2008. Balancing PPE protection with comfort, fit & style. Prof. Saf. 53, 50.

Miller, E., Buys, L., 2008. Retrofitting commercial office buildings for sustainability: tenants' perspectives. J. Prop. Invest. Financ. 26, 552–561.

Miller, D., Doh, J.H., Panuwatwanich, K., Van Oers, N., 2015. The contribution of structural design to green building rating systems: an industry perspective and comparison of life cycle energy considerations. Sustain. Cities Soc. 16, 39–48.

Mills, J.H., Dye, K., Mills, A.J., 2009. Understanding Organizational Change. Routledge, New York, NY.

Ministry of the Environment and Water Resources & Ministry of National Development, 2014. Sustainable Singapore Blueprint 2015. Ministry of the Environment and Water Resources and Ministry of National Development, Singapore.

Mojahed, S., Aghazadeh, F., 2008. Major factors influencing productivity of water and wastewater treatment plant construction: evidence from the deep south USA. Int. J. Proj. Manag. 26, 195–202.

Mokhlesian, S., Holmén, M., 2012. Business model changes and green construction processes. Constr. Manag. Econ. 30, 761–775.

Molenaar, K., Sobin, N., Gransberg, D., Mccuen, T., Korkmaz, S., Horman, M., 2009. Sustainable, high performance projects and project delivery methods: a state-of-practice report. In: White Paper for the Design-Build Institute of America and the Charles Pankow Foundation.

Mollaoglu-Korkmaz, S., Swarup, L., Riley, D., 2013. Delivering sustainable, high-performance buildings: influence of project delivery methods on integration and project outcomes. J. Manag. Eng. 29, 71–78.

MOM, 2006. A Guide to the Workplace Safety and Health (Risk Management) Regulations. Ministry of Manpower, Singapore.

Moresco, J., 2009. 5 Major Risks to Green Building Market Growth. Gigaom. Available from:http://gigaom.com/2009/09/02/5-major-risks-to-green-building-market-growth/ (Accessed 17 October 2013).

Morris, A., Zuo, J., Wang, Y., Wang, J., 2018. Readiness for sustainable community: a case study of green star communities. J. Clean. Prod. 173, 308–317.

MTI, 2015. MTI Maintains 2015 GDP Growth Forecast at 2.0 to 4.0 Per Cent. Ministry of Trade and Industry Singapore, Singapore. Available from: https://www.singstat.gov.sg/docs/default-source/default-document-library/news/press_releases/gdp4q2014.pdf (Accessed 4 April 2016).

MTI, 2016. Construction Working Group Report: Executive Summary. Ministry of Trade and Industry Singapore (MTI), Singapore. Available from: https://www.mti.gov.sg/researchroom/documents/app.mti.gov.sg/data/pages/507/doc/erc_dom_mainreport_part%203.5.pdf (Accessed 4 April 2016).

Mu, S., Cheng, H., Chohr, M., Peng, W., 2014. Assessing risk management capability of contractors in subway projects in mainland China. Int. J. Proj. Manag. 32, 452–460.

Müller, R., Turner, J.R., 2007. Matching the project manager's leadership style to project type. Int. J. Proj. Manag. 25, 21–32.

Mullins, L.J., 2007. Management and Organisational Behaviour. Prentice Hall Financial Times, Harlow, England.

Müngen, U., Gürcanli, G.E., 2005. Fatal traffic accidents in the Turkish construction industry. Saf. Sci. 43, 299–322.

Murakami, S., Kawakubo, S., Asami, Y., Ikaga, T., Yamaguchi, N., Kaburagi, S., 2011. Development of a comprehensive city assessment tool: CASBEE-city. Build. Res. Inf. 39, 195–210.

Mustapha, F., Naoum, S., 1998. Factors influencing the effectiveness of construction site managers. Int. J. Proj. Manag. 16, 1–8.

Nair, S., 2016. Singapore Budget 2016: Foreign Worker Levy Increase Deferred in Marine and Process Sectors. Available from: http://www.straitstimes.com/business/economy/singapore-budget-2016-foreign-worker-levy-increase-deferred-in-marine-and-process (Accessed 23 August 2016).

Nalewaik, A., Venters, V., 2010. Cost benefits of building green. IEEE Eng. Manag. Rev. 38, 77–87.

Newell, G., Sieracki, K., 2009. Global Trends in Real Estate Finance. Wiley, New York, United States.

Ng, A., Loosemore, M., 2007. Risk allocation in the private provision of public infrastructure. Int. J. Proj. Manag. 25, 66–76.

Ning, Y., Ling, F.Y.Y., Teo, A.C.Y., 2013. Driving forces behind and barriers to relational transaction practices in public construction projects. J. Prof. Issues Eng. Educ. Pract. 140.

Nofera, W., Korkmaz, S., 2010. In: Design process integration for sustainable, high performance buildings.Engineering Project Organizations Conference, South Lake Tahoe, CA.

Norman, G., 2010. Likert scales, levels of measurement and the "laws" of statistics. Adv. Health Sci. Educ. 15, 625–632.

Northouse, P.G., 2013. Leadership: Theory and Practice. Sage Publications, Thousand Oaks, CA.

Norušis, M.J., 2006. SPSS 14.0 Guide to Data Analysis. Prentice Hall, Upper Saddle River, NJ.

Nunnally, J.C., 1975. The study of change in evaluation research: principles con-concerning measurement, experimental design and analysis. In: Streuning, E.L., Guttentag, M. (Eds.), Handbook of Evaluation Research. Sage, Beverly Hills, CA.

Nunnally, J.C., 1978. Psychometric Theory. McGraw-Hill, New York, NY.

Nunnally, J.C., Bernstein, I.H., Berge, J.M.T., 1967. Psychometric Theory. JSTOR, New York, United States.

O'brien, R.M., 1979. The use of Pearson's R with ordinal data. Am. Sociol. Rev. 44, 851–857.

Odom, J.D., Scott, R., Dubose, G.H., 2008. Hidden Risks of Green Buildings. Contracting Business.

Odusami, K., 2002. Perceptions of construction professionals concerning important skills of effective project leaders. J. Manag. Eng. 18, 61–67.

Odusami, K.T., Iyagba, R.R.O., Omirin, M.M., 2003. The relationship between project leadership, team composition and construction project performance in Nigeria. Int. J. Proj. Manag. 21, 519–527.

OECD, 2015a. Mapping Channels to Mobilise Institutional Investment in Sustainable Energy. Available from: http://www.oecd-ilibrary.org/environment/mapping-channels-to-mobilise-institutional-investment-in-sustainable-energy_9789264224582-en (Accessed 15 June 2017).

OECD, 2015b. Overcoming Barriers to International Investment in Clean Energy. Available from: http://www.oecd-ilibrary.org/environment/overcoming-barriers-to-international-investment-in-clean-energy_9789264227064-en (Accessed 15 June 2017).

OECD, 2016a. Green Investment Banks: Scaling Up Private Investment in Low-Carbon, Climate-Resilient Infrastructure. Available from: http://www.oecd-ilibrary.org/finance-and-investment/green-investment-banks_9789264245129-en (Accessed 15 June 2017).

OECD, 2016b. OECD Centre on Green Finance and Investment. Available from: http://www.oecd.org/cgfi/brochure-oecd-centre-on-green-finance-and-investment.pdf (Accessed 15 June 2017).

OECD, 2016c. OECD Clean Energy Investment Policy Review of Jordan. Available from: http://www.oecd-ilibrary.org/finance-and-investment/oecd-clean-energy-investment-policy-review-of-jordan_9789264266551-en (Accessed 15 June 2017).

OECD, 2017. Mobilising Bond Markets for a Low-Carbon Transition. Available from: http://www.oecd-ilibrary.org/environment/mobilising-bond-markets-for-a-low-carbon-transition_9789264272323-en;jsessionid=1m2dao98rf5rd.x-oecd-live-02 (Accessed 15 June 2017).

Office of Energy Efficiency & Renewable Energy, 2017a. About the Energy Efficiency and Conservation Block Grant Program. Available from: https://energy.gov/eere/wipo/about-energy-efficiency-and-conservation-block-grant-program (Accessed 16 July 2017).

Office of Energy Efficiency & Renewable Energy, 2017b. Weatherization Assistance Program. Available from: https://energy.gov/eere/wipo/weatherization-assistance-program-1 (Accessed 16 July 2017).

Ofori, G., Kien, H.L., 2004. Translating Singapore architects' environmental awareness into decision making. Build. Res. Inf. 32, 27–37.

Ogunlana, S.O., Promkuntong, K., Jearkjirm, V., 1996. Construction delays in a fast-growing economy: comparing Thailand with other economies. Int. J. Proj. Manag. 14, 37–45.

Ökmen, Ö., Öztaş, A., 2008. Construction project network evaluation with correlated schedule risk analysis model. J. Constr. Eng. Manag. 134, 49–63.

Olomolaiye, P.O., Ogunlana, S.O., 1989. An evaluation of production outputs in key building trades in Nigeria. Constr. Manag. Econ. 7, 75–86.

Olomolaiye, P.O., Wahab, K.A., Price, D.F., 1987. Problems influencing craftsmen's productivity in Nigeria. Build. Environ. 22, 317–323.

Olubunmi, O.A., Xia, P.B., Skitmore, M., 2016. Green building incentives: a review. Renew. Sust. Energ. Rev. 59, 1611–1621.

Ong, Y.H., 2013. Green Building Retrofit Projects: Critical Success Factors, Barriers and Solutions (Undergraduate). National University of Singapore.

Ong, Y., 2014. Singapore's phantom workers. J. Contemp. Asia 44, 443–463.

Onkal-Engin, G., Demir, I., Hiz, H., 2004. Assessment of urban air quality in Istanbul using fuzzy synthetic evaluation. Atmos. Environ. 38, 3809–3815.

Ott, R.L., Longnecker, M., 2001. An Introduction to Statistical Methods and Data Analysis. Pacific Grove, Duxbury, MA.

Ott, R.L., Longnecker, M., 2010. An Introduction to Statistical Methods and Data Analysis. Brooks/Cole, Cengage Learning, Belmont, CA.

Ott, R.L., Longnecker, M.T., 2015. An Introduction to Statistical Methods and Data Analysis. Cengage Learning, Boston, United States.

Ozorhon, B., Arditi, D., Dikmen, I., Birgonul, M.T., 2008. Implications of culture in the performance of international construction joint ventures. J. Constr. Eng. Manag. 134, 361–370.

Palaneeswaran, E., Love, P.E.D., Kumaraswamy, M.M., Ng, T.S.T., 2008. Mapping rework causes and effects using artificial neural networks. Build. Res. Inf. 36, 450–465.

Pallant, J., 2010. SPSS Survival Manual: A Step by Step Guide to Data Analysis Using SPSS. McGraw-Hill, Maidenhead, UK.

Panthi, K., Ahmed, S.M., Ogunlana, S.O., 2009. Contingency estimation for construction projects through risk analysis. Int. J. Constr. Educ. Res. 5, 79–94.

Park, J., Yoon, J., Kim, K.H., 2017. Critical review of the material criteria of building sustainability assessment tools. Sustainability (Switzerland) 9, 186.

Patel, D.A., Jha, K.N., 2015. Neural network model for the prediction of safe work behavior in construction projects. J. Constr. Eng. Manag. 141 (1), 04014066.

Patil, V.H., Singh, S.N., Mishra, S., Donavan, D.T., 2007. Parallel Analysis Engine to Aid Determining Number of Factors to Retain.

Patlitzianas, K.D., Christos, K., 2011. Sustainable energy investments in Hellenic urban areas: examining modern financial mechanisms. Renew. Sustain. Energy Rev. 15, 5186–5193.

Patton, M.Q., 2005. Qualitative Research. Wiley Online Library, New York, United States.

Pearce, A., Kleiner, B., 2013. The Safety and Health of Construction Workers on Green Projects. Virginia Polytechnic Institute and State University.

Pearce, A., Ahn, Y.H., HanmiGlobal, 2012. Sustainable Buildings and Infrastructure. Routledge, New York.

Pedrycz, W., Ekel, P., Parreiras, R., 2011. Fuzzy Multicriteria Decision-Making: Models, Methods and Applications. John Wiley & Sons, Chichester, England.

Penny, J., 2012. The Cost of Green Buildings. Available from: http://www.buildings.com/article-details/articleid/13745/title/the-cost-of-green-buildings.aspx (Accessed 29 August 2013).

Perlman, A., Sacks, R., Barak, R., 2014. Hazard recognition and risk perception in construction. Saf. Sci. 64, 22–31.

Pinto, J.K., Slevin, D.P., 1987. Critical factors in successful project implementation. IEEE Trans. Eng. Manag. EM-34 (1), 22–27.

Pitt, V., 2013. Majority of Retrofit Projects Suffer Cost Overruns. Building. Available from: http://www.building.co.uk/majority-of-retrofit-projects-suffer-cost-overruns/5051213.article (Accessed 17 August 2013).

PMI, 2008. A Guide to the Project Management Body of Knowledge. Project Management Institute, Newtown Square, PA.

Polzin, F., 2017. Mobilizing private finance for low-carbon innovation—a systematic review of barriers and solutions. Renew. Sust. Energ. Rev. 77, 525–535.

Popovec, J., 2006. Incentives assuage investor fears over green premiums. Natl. Real Estate Invest. 48, 31.

Project Management Insititute (PMI), 2013. Project Management Body of Knowledge (PMBOK® Guide). Project Management Insititute, Inc., USA.

Project Management Institute, 2008. A Guide to the Project Management Body of Knowledge (PMBOK® Guide). Project Management Institute, USA.

Qin, X., Mo, Y., Jing, L., 2016. Risk perceptions of the life-cycle of green buildings in China. J. Clean. Prod. 126, 148–158.

Rajendran, S., Gambatese, J.A., 2009. Development and initial validation of sustainable construction safety and health rating system. J. Constr. Eng. Manag. 135, 1067–1075.

Rajendran, S., Gambatese, J.A., Behm, M.G., 2009. Impact of green building design and construction on worker safety and health. J. Constr. Eng. Manag. 135, 1058–1066.

Ranaweera, R., Crawford, R.H., 2010. Using early-stage assessment to reduce the financial risks and perceived barriers of sustainable buildings. J. Green Build. 5, 129–146.

Rastogi, A., Choi, J.K., Hong, T., Lee, M., 2017. Impact of different leed versions for green building certification and energy efficiency rating system: a multifamily midrise case study. Appl. Energy 205, 732–740.

Rebeiz, K.S., 2012. Public-private partnership risk factors in emerging countries: boot illustrative case study. J. Manag. Eng. 28, 421–428.

Rechenthin, D., 2004. Project safety as a sustainable competitive advantage. J. Saf. Res. 35, 297–308.

Regency Centers Corp, 2014. Regency Centers Sells USD 250 Million 'Green Bonds' Due 2014. Available from: http://www.snl.com/irweblinkx/file.aspx?iid=10309i&fid=23674110 (Accessed 30 June 2017).

Reijnders, L., Van Roekel, A., 1999. Comprehensiveness and adequacy of tools for the environmental improvement of buildings. J. Clean. Prod. 7, 221–225.

Retzlaff, R.C., 2008. Green building assessment systems: a framework and comparison for planners. J. Am. Plan. Assoc. 74, 505–519.

Retzlaff, R.C., 2009. Green buildings and building assessment systems: a new area of interest for planners. J. Plan. Lit. 24, 3–21.

Riemer, J.W., 1976. "Mistakes at work" the social organization of error in building construction work. Soc. Probl., 255–267.

Robbins, S.P., 2003. Organizational Behavior. Prentice Hall, Upper Saddle River, NJ.

Robbins, S.P., Coulter, M., 2007. Management. Pearson Prentice Hall, Upper Saddle River, NJ.

Robichaud, L.B., Anantatmula, V.S., 2010. Greening project management practices for sustainable construction. J. Manag. Eng. 27, 48–57.

Robichaud, L.B., Anantatmula, V.S., 2011. Greening project management practices for sustainable construction. J. Manag. Eng. 27, 48–57.

Roumboutsos, A., Anagnostopoulos, K.P., 2008. Public private partnership projects in Greece: risk ranking and preferred risk allocation. Constr. Manag. Econ. 26, 751–763.

Rowlinson, S., Yunyanjia, A., Li, B., Chuanjingju, C., 2014. Management of climatic heat stress risk in construction: a review of practices, methodologies, and future research. Accid. Anal. Prev. 66, 187–198.

Rupp, R.F., Vásquez, N.G., Lamberts, R., 2015. A review of human thermal comfort in the built environment. Energy Build. 105, 178–205.

Russell, J.S., Jaselskis, E.J., Lawrence, S.P., 1997. Continuous assessment of project performance. J. Constr. Eng. Manag. 123, 64–71.

Rybkowski, Z.K., Shepley, M.M., Bryant, J.A., Skelhorn, C., Amato, A., Kalantari, S., 2017. Facility management in Qatar: current state, perceptions and recommendations. Facilities 35, 335–355.

Sahamir, S.R., Zakaria, R., 2013. Green assessment criteria for public hospital building development in Malaysia. Procedia Environ. Sci. 20, 106–115.

Santos, J.R.A., 1999. Cronbach's alpha: a tool for assessing the reliability of scales. J. Ext. 37, 1–5.

Sanvido, V.E., Riggs, L.S., 1991. Managing Retrofit Projects. Construction Industry Institute, Austin, TX.

Schoonover, T., Bonauto, D., Silverstein, B., Adams, D., Clark, R., 2010. Prioritizing prevention opportunities in the Washington State construction industry, 2003–2007. J. Saf. Res. 41, 197–202.

Schwartz, Y., Raslan, R., 2013. Variations in results of building energy simulation tools, and their impact on breeam and Leed ratings: a case study. Energy Build. 62, 350–359.

Schweber, L., 2013. The effect of Breeam on clients and construction professionals. Build. Res. Inf. 41, 129–145.

Seinre, E., Kurnitski, J., Voll, H., 2014. Building sustainability objective assessment in Estonian context and a comparative evaluation with Leed and Breeam. Build. Environ. 82, 110–120.

Shan, M., Chan, A., Le, Y., Xia, B., Hu, Y., 2015a. Measuring corruption in public construction projects in China. J. Prof. Issues Eng. Educ. Pract. 141. https://doi.org/10.1061/(Asce)Ei.1943-5541.0000241.

Shan, M., Chan, A.P., Le, Y., Hu, Y., 2015b. Investigating the effectiveness of response strategies for vulnerabilities to corruption in the Chinese public construction sector. Sci. Eng. Ethics 21, 683–705.

Shan, M., Chan, A.P., Le, Y., Hu, Y., Xia, B., 2016. Understanding collusive practices in Chinese construction projects. J. Prof. Issues Eng. Educ. Pract.

Shen, Q., Liu, G., 2003. Critical success factors for value management studies in construction. J. Constr. Eng. Manag. 129, 485–491.

Shen, L., Wu, G.W., Ng, C.S., 2001. Risk assessment for construction joint ventures in China. J. Constr. Eng. Manag. 127, 76–81.

Shi, Q., Zuo, J., Zillante, G., 2012. Exploring the management of sustainable construction at the programme level: a Chinese case study. Constr. Manag. Econ. 30, 425–440.

Shi, Q., Zuo, J., Huang, R., Huang, J., Pullen, S., 2013. Identifying the critical factors for green construction—an empirical study in China. Habitat Int. 40, 1–8.

Shi, Q., Yan, Y., Zuo, J., Yu, T., 2016. Objective conflicts in green buildings projects: a critical analysis. Build. Environ. 96, 107–117.

Shiers, D., Rapson, D., Roberts, C., Keeping, M., 2006. Sustainable construction: the development and evaluation of an environmental profiling system for construction products. Constr. Manag. Econ. 24, 1177–1184.

Siamanta, Z.C., 2017. Building a green economy of low carbon: the Greek post-crisis experience of photovoltaics and financial 'green grabbing'. J. Polit. Ecol. 24, 258–276.

Singh, A., Syal, M., Grady, S.C., Korkmaz, S., 2010. Effects of green buildings on employee health and productivity. Am. J. Public Health 100, 1665–1668.

Skorupka, D., 2008. Identification and initial risk assessment of construction projects in Poland. J. Manag. Eng. 24, 120–127.

Smallwood, J., 2006. The practice of construction management. Acta Struct. J. Phys. Dev. Sci. 13, 62–89.

Snell, C., Callahan, T., 2005. Building Green: A Complete How-to Guide to Alternative Building Methods: Earth Plaster, Straw Bale, Cordwood, Cob, Living Roofs. Lark Books, New York, United States.

Son, H., Lee, S., Kim, C., 2015. An empirical investigation of key pre-project planning practices affecting the cost performance of green building projects. Procedia Eng. 118, 37–41.

Soundarrajan, P., Vivek, N., 2016. Green finance for sustainable green economic growth in India. Agric. Econ. Zemed. Ekon. 62, 35–44.

Spillane, J.P., Oyedele, L.O., Von Meding, J., Jaiyeoba, B.E., Tijani, I.K., 2011. Confined site construction: a qualitative investigation of critical issues affecting management of health and safety. J. Civ. Eng. Constr. Technol. 2, 138–146.

SPRING Singapore, 2011. A Guide to Productivity Measurement. SPRING Singapore, Singapore.

Statistics Solutions, 2017. Paired Sample T-Test. Available from: http://www.statisticssolutions.com/manova-analysis-paired-sample-t-test/ (Accessed 3 July 2017).

Stern, N.H., 2007. The Economics of Climate Change: The Stern Review. Cambridge University Press, Cambridge, United Kingdom.

Still, J., Orlet, H., Law, E., Wheeler, M., Pickens, H., 1997. Electrocution due to contact of industrial equipment with power lines. Burns 23, 573–575.

Stogdill, R.M., Coons, A.E., 1951. Leader Behavior: Its Description and Measurement. Ohio State University and Bureau of Business Research, Columbus, OH.

Streimikiene, D., 2016. Review of financial support from EU structural funds to sustainable energy in Baltic states. Renew. Sustain. Energy Rev. 58, 1027–1038.

Strielkowski, W., Štreimikienė, D., Bilan, Y., 2017. Network charging and residential tariffs: a case of household photovoltaics in the United Kingdom. Renew. Sust. Energ. Rev. 77, 461–473.

Sustainable Melbourne Fund, 2013. The #1 Barrier to Green Buildings Is Finance. Available from: http://sustainablemelbournefund.com.au/1-barrier-green-buildings-finance/ (Accessed 20 June 2017).

Sweis, G., Sweis, R., Abu Hammad, A., Shboul, A., 2008. Delays in Construction Projects: The Case of Jordan. Int. J. Proj. Manag. 26, 665–674.

Swuste, P., 2013. A 'normal accident'with a tower crane? an accident analysis conducted by the Dutch Safety Board. Saf. Sci. 57, 276–282.

Tagaza, E., Wilson, J., 2004. Green Buildings: Drivers and Barriers—Lessons Learned From Five Melbourne Developments. Report prepared for Building Commission by University of Melbourne and Business Outlook and Evaluation.

Tam, V.W., Fung, I.W., 2011. Tower crane safety in the construction industry: a Hong Kong study. Saf. Sci. 49, 208–215.

Tam, C.M., Zeng, S.X., Deng, Z.M., 2004. Identifying elements of poor construction safety management in China. Saf. Sci. 42, 569–586.

Tan, M., 2015. SEF Group's Prefab Hub Opens in Kaki Bukit. Singapore. Available from: http://www.businesstimes.com.sg/real-estate/sef-groups-prefab-hub-opens-in-kaki-bukit (Accessed 29 March 2016).

Tang, W., Qiang, M., Duffield, C.F., Young, D.M., Lu, Y., 2007. Risk management in the Chinese construction industry. J. Constr. Eng. Manag. 133, 944–956.

Tay, E., 2015. 2015 Guide to Singapore Government Funding and Incentives for the Environment. Available from: http://www.greenfuture.sg/2015/02/16/2015-guide-to-singapore-government-funding-and-incentives-for-the-environment/ (Accessed 23 July 2017).

Taylor, D.H., Hensey, M., 1990. Financial issues in engineering management: interview. J. Manag. Eng. 6, 157–161.

Teo, P.J.P., 2008. Adjudication: Singapore perspective. J. Prof. Issues Eng. Educ. Pract. 134, 224–230.

Tett, R.P., Guterman, H.A., Bleier, A., Murphy, P.J., 2000. Development and content validation of a "hyperdimensional" taxonomy of managerial competence. Hum. Perform. 13, 205–251.

Thatcher, A., Milner, K., 2014. Changes in productivity, psychological wellbeing and physical wellbeing from working in a 'green' building. Work 49, 381–393.

The Straits Times, 2016. Construction Company Singbuild Fined $7,000 for Illegal Dumping and Pollution. Available from: http://www.straitstimes.com/singapore/courts-crime/construction-company-singbuild-fined-7000-for-illegal-dumping-and-pollution (Accessed 18 May 2016).

The World Bank, 2015. What Are Green Bonds? The World Bank, Washington, DC.

The World Bank, 2017. Implementation Competion and Results Report on a Grant From the Global Environment Facility Trust Fund in the Amount of USD 6,163,636 to China for the Sino-Singapore Tianjin Eco-City Project. .

Thomas, H.R., Napolitan, C.L., 1995. Quantitative effects of construction changes on labor productivity. J. Constr. Eng. Manag. 121, 290–296.

Thomas, H.R., Riley, D.R., Sanvido, V.E., 1999. Loss of labor productivity due to delivery methods and weather. J. Constr. Eng. Manag. 125, 39–46.

Tobias, L., Malachite, L., 2008. Toward Sustainable Financing and Strong Markets for Green Building: US Green Building Finance Review. Available from: http://www3.cec.org/islandora/en/item/2328-paper-2b-toward-sustainable-financing-and-strong-markets-green-building-en.pdf (Accessed 21 June 2017).

Tollin, H.M., 2011. Green building risks: it's not easy being green. Environ. Claims J. 23, 199–213.

Toroghi, S.H., Nguyen, T.H., Jacobs, F., 2016. Automated green building rating system for building designs. J. Archit. Eng.. 22.

Trading Economics, 2016. Singapore-Inflation Rate. Available from: http://zh.tradingeconomics.com/singapore/inflation-cpi (Accessed 20 April 2016).

Turner, R., 1999. Handbook of Project-Based Management: Improving the Processes of Achieving Strategic Objectives. McGraw-Hill, London.

Turner, R., 2009. Handbook of Project-Based Management: Leading Strategic Change in Organizations. McGraw-Hill, London.

Turner, J.R., Müller, R., 2005. The project manager's leadership style as a success factor on projects: a literature review. Proj. Manag. J. 36, 49–61.

U.S. Green Building Council, 2017. Leed V4 for Building Design and Construction. Available from: https://www.usgbc.org/sites/default/files/leed%20v4%20bdc_07.8.17_current.pdf (Accessed 29 September 2017).

Uher, T.E., Brand, M.C., 2008. Claimants' view of the performance of adjudication in New South Wales. Eng. Constr. Archit. Manag. 15, 470–484.

UNEP, 2001. Energy and Cities: Sustainable Building and Construction. United Nations. Available from: http://www.unep.or.jp/ietc/focus/energycities1.asp (Accessed 18 December 2016).

UNEP, 2009. Building and Climate Change: Summary for Decision-Makers. United Nations Environment Programme, Paris, France.

UNEP Finance Initiative, 2010. Green Buildings and the Finance Sector. David Gardiner & Associates, LLC, Washington DC.

UNEP Finance Initiative, 2017. About United Nations Environment Programme-Finance Initiative. Available from: http://www.unepfi.org/about/ (Accessed 15 June 2017).

United Nations Environment Programme (UNEP), 2011. Sustainable Buildings and Climate Initiatives. United Nations Environmental Programme, Paris. Available from: http://www.unep.org/sbci/pdfs/sbci_2pager_eversion_feb2011.pdf (Accessed 12 February 2015).

United States Green Building Council, 2003. Building Momentum: National Trends and Prospects for High-Performance Green Buildings: Based on the April 2002 Green Building Roundtable and Prepared for the US Senate Committee on Environment and Public Works. US Green Building Council.

United States Green Building Council (USGBC), 2007. A National Green Building Research Agenda. US Green Building Council, United States.

United States Green Building Council (USGBC), 2009. Leed for New Construction and Major Renovations. United States Green Building Council, Washington, DC.

Vaughan, A., 2012. Green Deal: Households to Get Up to £1,000 for Insulation. Available from: https://www.theguardian.com/environment/2012/oct/19/green-deal-cashback-energy-efficiency (Accessed 14 June 2017).

Von Paumgartten, P., 2003. The business case for high performance green buildings: sustainability and its financial impact. J. Facil. Manag. 2, 26–34.

Walter, L., 2011. "Green" Construction Workers May Face Additional Safety Risks. Available from: http://libproxy1.nus.edu.sg/login?url=http://search.proquest.com/docview/906955972?accountid=13876 (Accessed 16 August 2014).

Wang, S.Q., Dulaimi, M.F., Aguria, M.Y., 2004. Risk management framework for construction projects in developing countries. Constr. Manag. Econ. 22, 237–252.

Wang, T., Wang, S., Zhang, L., Huang, Z., Li, Y., 2015. A major infrastructure risk-assessment framework: application to a cross-sea route project in China. Int. J. Proj. Manag. 34 (7), 1403–1415.

Weber, B., Alfen, H.W., Staub-Bisang, M., 2016. Infrastructure as an Asset Class: Investment Strategy, Sustainability, Project Finance and PPP. John Wiley & Sons, New York, United States.

Wedding, G.C., Crawford-Brown, D., 2008. Improving the link between the leed green building label and a building's energy-related environmental metrics. J. Green Build. 3, 85–105.

Wibowo, A., Kochendoerfer, B., 2011. Selecting BOT/PPP infrastructure projects for government guarantee portfolio under conditions of budget and risk in the Indonesian context. J. Constr. Eng. Manag. 137, 512–522.

Wibowo, A., Kochendörfer, B., 2005. Financial risk analysis of project finance in Indonesian toll roads. J. Constr. Eng. Manag. 131, 963–972.

Wibowo, A., Mohamed, S., 2010. Risk criticality and allocation in privatised water supply projects in Indonesia. Int. J. Proj. Manag. 28, 504–513.

Wilkins, J.R., 2011. Construction workers' perceptions of health and safety training programmes. Constr. Manag. Econ. 29, 1017–1026.

Williams Jr., Q., Ochsner, M., Marshall, E., Kimmel, L., Martino, C., 2010. The impact of a peer-led participatory health and safety training program for latino day laborers in construction. J. Saf. Res. 41, 253–261.

Winston, A.S., 2009. Green Recovery. Harvard Business Press, Boston, MA.

Wong, S.-C., Abe, N., 2014. Stakeholders' perspectives of a building environmental assessment method: the case of CASBEE. Build. Environ. 82, 502–516.

Wong, J.K.-W., Kuan, K.-L., 2014. Implementing 'beam plus' for BIM-based sustainability analysis. Autom. Constr. 44, 163–175.

Wong, J.K.W., Zhou, J., 2015. Enhancing environmental sustainability over building life cycles through green BIM: a review. Autom. Constr. 57, 156–165.

Wong, N.H., Tay, S.F., Wong, R., Ong, C.L., Sia, A., 2003. Life cycle cost analysis of rooftop gardens in Singapore. Build. Environ. 38, 499–509.

World Bank Treasury, 2017. What Are Green Bonds? Available from: http://treasury.worldbank.org/cmd/htm/worldbankgreenbonds.html (Accessed 20 June 2017).

World Health Organization, 2016. International Statistical Classification of Diseases and Related Health Problems 10th Revision (ICD-10)-Who Version for 2016. Available from: http://apps.who.int/classifications/icd10/browse/2016/en#/xx (Accessed 10 March 2016).

WorldGBC, 2013. The Business Case for Green Building—A Review of the Costs and Benefits for Developers, Investors and Occupants. Available from: http://www.worldgbc.org/files/1513/6608/0674/business_case_for_green_building_report_web_2013-04-11.pdf (Accessed 22 December 2014).

WSHC, 2010. Implementing WSH 2018 for the Construction Sector in Singapore. Singapore. Available from: https://www.wshc.sg/wps/themes/html/upload/cms/file/wsh%20construction%20sectoral%20final_lowres(1).pdf (Accessed 20 May 2014).

WSHC, 2013. Workplace Safety and Health Report, 2013. National Statistics, Singapore. Available from: https://www.wshc.sg/wps/pa_infostop/download/wsh2013.pdf?folder=is2014041000789&file=wsh2013.pdf (Accessed 20 May 2014).

Wu, P., Feng, Y., 2014. Identification of non-value adding activities in precast concrete production to achieve low-carbon production. Archit. Sci. Rev. 57, 105–113.

Wu, W., Issa, R.R.A., 2014. BIM execution planning in green building projects: Leed as a use case. J. Manag. Eng. 31 (1), 4014007.

Wu, W., Issa Raja, R.A., 2015. BIM execution planning in green building projects: Leed as a use case. J. Manag. Eng. 31.

Wu, P., Low, S.P., 2010. Project management and green buildings: lessons from the rating systems. J. Prof. Issues Eng. Educ. Pract. 136, 64–70.

Wu, P., Low, S.P., 2012. Lean management and low carbon emissions in precast concrete factories in Singapore. J. Archit. Eng. 18, 176–186.

Wu, P., Low, S.P., 2014. Barriers to achieving green precast concrete stock management—a survey of current stock management practices in Singapore. Int. J. Constr. Manag. 14, 78–89.

Wu, C.H., Hsieh, T.Y., Cheng, W.L., 2005. Statistical analysis of causes for design change in highway construction on Taiwan. Int. J. Proj. Manag. 23, 554–563.

Wu, W., Yang, H., Li, Q., Chew, D., 2013. An integrated information management model for proactive prevention of struck-by-falling-object accidents on construction sites. Autom. Constr. 34, 67–74.

Wu, P., Xia, B., Pienaar, J., Zhao, X., 2014a. The past, present and future of carbon labelling for construction materials—a review. Build. Environ. 77, 160–168.

Wu, P., Xia, B., Zhao, X., 2014b. The importance of use and end-of-life phases to the life cycle greenhouse gas (GHG) emissions of concrete—a review. Renew. Sust. Energ. Rev. 37, 360–369.

Wu, P., Feng, Y., Pienaar, J., Xia, B., 2015. A review of benchmarking in carbon labelling schemes for building materials. J. Clean. Prod. 109, 108–117.

Wu, Z., Shen, L., Yu, A.T.W., Zhang, X., 2016. A comparative analysis of waste management requirements between five green building rating systems for new residential buildings. J. Clean. Prod. 112, 895–902.

Wu, P., Song, Y., Shou, W., Chi, H., Chong, H.-Y., Sutrisna, M., 2017. A comprehensive analysis of the credits obtained by Leed 2009 certified green buildings. Renew. Sust. Energ. Rev. 68, 370–379.

Xenidis, Y., Angelides, D., 2005. The financial risks in build-operate-transfer projects. Constr. Manag. Econ. 23, 431–441.

Xia, B., Chan, A.P., 2011. Investigation of barriers to entry into the design-build market in the People's Republic of China. J. Constr. Eng. Manag. 138, 120–127.

Xia, B., Chan, A.P.C., Yeung, J.F.Y., 2011. Developing a fuzzy multicriteria decision-making model for selecting design-build operational variations. J. Constr. Eng. Manag. 137, 1176–1184.

Xu, Y., Chan, A.P.C., Yeung, J.F.Y., 2010a. Developing a fuzzy risk allocation model for PPP projects in China. J. Constr. Eng. Manag. 136, 894–903.

Xu, Y., Yeung, J.F.Y., Chan, A.P.C., Chan, D.W.M., Wang, S.Q., Ke, Y., 2010b. Developing a risk assessment model for PPP projects in China—a fuzzy synthetic evaluation approach. Autom. Constr. 19, 929–943.

Yamaguchi, R., Managi, S., 2017. New financing for sustainable development: the case for NNP- or inclusive wealth-linked bonds. J. Environ. Dev. 26, 214–239.

Yang, R.J., Zou, P.X.W., 2014. Stakeholder-associated risks and their interactions in complex green building projects: a social network model. Build. Environ. 73, 208–222.

Yang, L.-R., Huang, C.-F., Wu, K.-S., 2011. The association among project manager's leadership style, teamwork and project success. Int. J. Proj. Manag. 29, 258–267.

Yang, R.J., Zou, P.X.W., Wang, J., 2016. Modelling stakeholder-associated risk networks in green building projects. Int. J. Proj. Manag. 34, 66–81.

Ye, L., Cheng, Z., Wang, Q., Lin, W., Ren, F., 2013. Overview on green building label in China. Renew. Energy 53, 220–229.

Ye, L., Cheng, Z., Wang, Q., Lin, H., Lin, C., Liu, B., 2015. Developments of green building standards in China. Renew. Energy 73, 115–122.

Yi, W., Chan, A.P., 2013. Optimizing work–rest schedule for construction rebar workers in hot and humid environment. Build. Environ. 61, 104–113.

Yik, F., Burnett, J., Prescott, I., 2001. Predicting air-conditioning energy consumption of a group of buildings using different heat rejection methods. Energy Build. 33, 151–166.

Yoon, S.W., Lee, D.K., 2003. The development of the evaluation model of climate changes and air pollution for sustainability of cities in Korea. Landsc. Urban Plan. 63, 145–160.

Yu, W., Li, B., Yang, X., Wang, Q., 2015. A development of a rating method and weighting system for green store buildings in China. Renew. Energy 73, 123–129.

Yuan, X., Zuo, J., 2013. A critical assessment of the higher education for sustainable development from students' perspectives—a Chinese study. J. Clean. Prod. 48, 108–115.

Yuan, X., Zuo, J., Huisingh, D., 2015. Social acceptance of wind power: a case study of Shandong province, China. J. Clean. Prod. 92, 168–178.

Yudelson, J., 2010a. The Green Building Revolution. Island Press, Washington, D.C., United States.

Yudelson, J., 2010b. Greening Existing Buildings. McGraw-Hill, New York, NY.

Yudelson, J., Fedrizzi, S.R., 2010. The Green Building Revolution. Island Press, Washington, D.C., United States.

Zaneldin, E., Hegazy, T., Grierson, D., 2001. Improving design coordination for building projects. II: a collaborative system. J. Constr. Eng. Manag. 127, 330–336.

Zhan, C.J., De Jong, M., 2017. Financing Sino-Singapore Tianjin eco-city: what lessons can be drawn for other large-scale sustainable city-projects? Sustainability 9, 201.

Zhang, X., 2014. Paradigm shift toward sustainable commercial project development in China. Habitat Int. 42, 186–192.

Zhang, X., Shen, L., Wu, Y., 2011. Green strategy for gaining competitive advantage in housing development: a China study. J. Clean. Prod. 19, 157–167.

Zhang, D., Haas, C.T., Goodrum, P.M., Caldas, C.H., Granger, R., 2012. Construction small-projects rework reduction for capital facilities. J. Constr. Eng. Manag. 138, 1377–1385.

Zhang, R.P., Lingard, H., Nevin, S., 2015. Development and validation of a multilevel safety climate measurement tool in the construction industry. Constr. Manag. Econ. 33 (10), 818–839.

Zhang, Y., Wang, J., Hu, F., Wang, Y., 2017. Comparison of evaluation standards for green building in China, Britain, United States. Renew. Sust. Energ. Rev. 68, 262–271.

Zhao, X., 2017. A scientometric review of global BIM research: analysis and visualization. Autom. Constr. 80, 37–47.

Zhao, X., Hwang, B.-G., Yu, G.S., 2013a. Identifying the critical risks in underground rail international construction joint ventures: case study of Singapore. Int. J. Proj. Manag. 31, 554–566.

Zhao, X., Hwang, B.G., Low, S.P., 2013b. Critical success factors for enterprise risk management in Chinese construction companies. Constr. Manag. Econ. 31, 1199–1214.

Zhao, X., Hwang, B.G., Low, S.P., 2013c. Developing fuzzy enterprise risk management maturity model for construction firms. J. Constr. Eng. Manag. 139, 1179–1189.

Zhao, X., Hwang, B.G., Low, S.P., 2014a. Investigating enterprise risk management maturity in construction firms. J. Constr. Eng. Manag. 140.

Zhao, X., Hwang, B.G., Phng, W., 2014b. Construction project risk management in Singapore: resources, effectiveness, impact, and understanding. KSCE J. Civ. Eng. 18, 27–36.

Zhao, X., Hwang, B.G., Low, S.P., Wu, P., 2015. Reducing hindrances to enterprise risk management implementation in construction firms. J. Constr. Eng. Manag. 141.

Zhao, X., Hwang, B.G., Gao, Y., 2016. A fuzzy synthetic evaluation approach for risk assessment: a case of Singapore's green projects. J. Clean. Prod. 115, 203–213.

Zhi, H., 1995. Risk management for overseas construction projects. Int. J. Proj. Manag. 13, 231–237.

Zhou, W., Whyte, J., Sacks, R., 2012. Construction safety and digital design: a review. Autom. Constr. 22, 102–111.

Zhou, Z., Goh, Y.M., Li, Q., 2015. Overview and analysis of safety management studies in the construction industry. Saf. Sci. 72, 337–350.

Zhu, Y., Song, L., Damiens, J., 2013. State of the art of green building standards and labelling system development in China. Int. J. Sustain. Build. Technol. Urban Dev. 4, 178–184.

Zimmermann, H.J., 2001. Fuzzy Set Theory and Its Applications. Kluwer Academic Publishers, Boston, MA.

Zou, P.X.W., Couani, P., 2012. Managing risks in green building supply chain. Archit. Eng. Des. Manag. 8, 143–158.

Zou, P.X.W., Li, J., 2010. Risk identification and assessment in subway projects: case study of Nanjing subway line 2. Constr. Manag. Econ. 28, 1219–1238.

Zou, P.X., Zhang, G., Wang, J., 2007. Understanding the Key Risks in Construction Projects in China. Int. J. Proj. Manag. 25, 601–614.

Zumbo, B.D., Zimmerman, D.W., 1993. Is the selection of statistical methods governed by level of measurement? Can. Psychol. 34, 390.

Zuo, J., Zhao, Z.-Y., 2014. Green building research–current status and future agenda: a review. Renew. Sust. Energ. Rev. 30, 271–281.

Zuo, J., Read, B., Pullen, S., Shi, Q., 2013. Carbon-neutral commercial building development. J. Manag. Eng. 29, 95–102.

Zuo, J., Pullen, S., Palmer, J., Bennetts, H., Chileshe, N., Ma, T., 2015. Impacts of heat waves and corresponding measures: a review. J. Clean. Prod. 92, 1–12.

Zuo, J., Xia, B., Chen, Q., Pullen, S., Skitmore, M., 2016. Green building rating for office buildings—lessons learned. J. Green Build. 11, 131–146.

Zurich Services Corporation, 2011. Green Building: What Are the Risks? Zurich Services Corporation, Schaumburg, IL. Available from: http://hpd.zurichna.com/whitepaper/zurich-re-advisen-green-building.pdf (Accessed 7 February 2014).

Index

Note: Page numbers followed by *f* indicate figures, and *t* indicate tables.

G

T

U

W